Communications and Control Engineering

Springer
*London
Berlin
Heidelberg
New York
Barcelona
Budapest
Hong Kong
Milan
Paris
Santa Clara
Singapore
Tokyo*

Published titles include:

Sampled-Data Control Systems
J. Ackermann

The Riccatti Equation
S. Bittanti, A.J. Laub and J.C. Willems (Eds)

Sliding Modes in Control Optimization
V.I. Utkin

Fundamentals of Robotics
M. Vukobratović

Parametrizations in Control, Estimation and Filtering Problems: Accuracy Aspects
M. Gevers and G. Li

Parallel Algorithms for Optimal Control of Large Scale Linear Systems
Zoran Gajić and Xuemin Shen

Loop Transfer Recovery: Analysis and Design
A. Saberi, B.M. Chen and P. Sannuti

Markov Chains and Stochastic Stability
S.P. Meyn and R.L. Tweedie

Robust Control: Systems with Uncertain Physical Parameters
J. Ackermann in co-operation with A. Bartlett, D. Kaesbauer, W. Sienel and R. Steinhauser

Optimization and Dynamical Systems
U. Helmke and J.B. Moore

Optimal Sampled-Data Control Systems
Tongwen Chen and Bruce Francis

Nonlinear Control Systems (3rd edition)
Alberto Isidori

Theory of Robot Control
C. Canudas de Wit, B. Siciliano and G. Bastin (Eds)

Fundamental Limitations in Filtering and Control
María M. Seron, Julio Braslavsky and Graham C. Goodwin

Constructive Nonlinear Control
R. Sepulchre, M. Janković and P.V. Kokotović

A Theory of Learning and Generalization
M. Vidyasagar

Adaptive Control
I.D. Landau, R Lozano and M. M'Saad

Miroslav Krstić and Hua Deng

Stabilization of Nonlinear Uncertain Systems

With 14 Figures

Springer

Miroslav Krstić
Hua Deng
University of California, San Diego, La Jolla, CA 92093-0411, USA

Series Editors
B.W. Dickinson • A. Fettweis • J.L. Massey • J.W. Modestino
E.D. Sontag • M. Thoma

ISBN 1-85233-020-1 Springer-Verlag Berlin Heidelberg New York

British Library Cataloguing in Publication Data
Krstic, Miroslav
  Stabilization of nonlinear uncertain systems. -
  (Communication and control engineering series)
  1.Nonlinear systems 2.Nonlinear control theory
  I.Title II.Deng, Hua
  629.8'36
ISBN 1852330201

Library of Congress Cataloging-in-Publication Data
Krstić, Miroslav.
   Stabilization of nonlinear uncertain systems / Miroslav Krstić and
   Hua Deng.
       p.   cm. -- (Communications and control engineering)
     Includes bibliographical references and index.
     ISBN 1-85233-020-1 (casebound)
     1. Feedback control systems. 2. Nonlinear control theory.
   3. Telecommunication systems. I. Deng, Hua, 1969- . II. Title.
   III. Series.
   TJ216.K77 1998
   629.8'3--dc21   98-3137

Apart from any fair dealing for the purposes of research or private study, or criticism or review, as permitted under the Copyright, Designs and Patents Act 1988, this publication may only be reproduced, stored or transmitted, in any form or by any means, with the prior permission in writing of the publishers, or in the case of reprographic reproduction in accordance with the terms of licences issued by the Copyright Licensing Agency. Enquiries concerning reproduction outside those terms should be sent to the publishers.

© Springer-Verlag London Limited 1998
Printed in Great Britain

The use of registered names, trademarks, etc. in this publication does not imply, even in the absence of a specific statement, that such names are exempt from the relevant laws and regulations and therefore free for general use.

The publisher makes no representation, express or implied, with regard to the accuracy of the information contained in this book and cannot accept any legal responsibility or liability for any errors or omissions that may be made.

Typesetting: Camera ready by authors
Printed and bound at the Athenæum Press Ltd., Gateshead, Tyne & Wear
69/3830-543210 Printed on acid-free paper

Preface

The subject of this monograph is global stabilization and optimal control of nonlinear systems with uncertain models. The book offers a unified view of (I) *deterministic disturbance attenuation*, (II) *stochastic control*, and (III) *adaptive control* for nonlinear systems. In order to provide the big picture, we take the approach of presenting the fundamentals, rather than giving a comprehensive coverage for the most general classes of systems.

The efforts in nonlinear control over the last few years have led to the level of generality in which conceptual solutions exist for all systems affine in control and the "disturbance," and constructive designs exist for a specific (but broad in terms of physical relevance) class of nonlinear systems. The intention of the book is to communicate this message to a large audience of researchers, students, engineers, and mathematicians, in the areas of robust and adaptive nonlinear control, nonlinear $\mathcal{H}_\infty$, stochastic nonlinear control (including risk-sensitive) and other related areas of control and dynamical systems theory.

While the book covers a broad area to which many people have contributed over the years to bring it to the current status, the results in the book are based on the authors' research in the last three years and includes dissertation work by the second author with the first author, her PhD advisor.

Parts I and II deal with systems described, respectively, by the differential equations

$$\dot{x} = f(x) + g_1(x)d + g_2(x)u$$
$$dx = f(x)dt + g_1(x)dw + g_2(x)udt$$

where u is control, $d(t)$ is a deterministic disturbance, and $w(t)$ is an independent Wiener process with $E\{dwdw^\mathrm{T}\} = \Sigma(t)\Sigma(t)^\mathrm{T}dt$. The most general problems solved for these systems are when $d(t)$ and $\Sigma(t)$ are functions whose bound is not known to the control designer or when $g_1(0) \neq 0$, which prevents the existence of an equilibrium at the origin in the presence of the disturbance.

In these cases, controllers are designed which guarantee that[1]

$$|x(t)| < \alpha \left(\sup_{[0,t]} |d(\tau)| \right)$$

$$P\left\{ |x(t)| < \alpha \left(\sup_{[0,t]} \left| \Sigma(\tau)\Sigma(\tau)^{\mathrm{T}} \right| \right) \right\} = 1$$

for all $t \geq 0$, and

$$\int_0^\infty \left[l(x) + \gamma_2(|u|) \right] dt \leq \int_0^\infty \gamma_1(|d|) \, dt$$

$$E\left\{ \int_0^\infty \left[l(x) + \gamma_2(|u|) \right] dt \right\} \leq \int_0^\infty \gamma_1\left(\left| \Sigma \Sigma^{\mathrm{T}} \right| \right) dt$$

where $\alpha, \gamma_1, \gamma_2$ are some continuous and strictly increasing functions vanishing at zero and l is a positive definite function. The book shows that the respective problems are equivalent and provides constructive solutions based on control Lyapunov functions. The focus is on inverse (rather than direct) optimality because it avoids Hamilton-Jacobi-Isaacs pde's and still results in Kalman-type stability margins.

Part III deals with parametric uncertainties. For example, for the system

$$\dot{x} = f(x) + g_1(x)\theta + g_2(x)u,$$

where θ is a constant unknown vector, it develops controllers which guarantee global stability, regulation of x to zero, and optimality with respect to a cost functional that involves integral penalty on x and u, and terminal penalty on the parameter estimation error $\theta - \hat{\theta}$. Since this type of optimality leaves open the question of asymptotic behavior of the parameter estimate $\hat{\theta}$, **Chapter 7** gives an in-depth geometric study of convergence via invariant manifold theory, a first study of this kind in the adaptive literature which shows that not every adaptive controller converges to a stabilizing frozen controller.

Chapter 8, the last chapter of the book, addresses a very different type of adaptive optimal control problem, the one in which the equilibrium is unknown and the objective is to optimize steady state performance. In this problem the system is not required to be affine in either the control or the parameter but the results are local and achieved via averaging and singular perturbations.

Chapters 1–6 require little background beyond basic linear systems theory (including stochastic systems). Lyapunov theory is included in the book, although not to substitute for a pedagogical text like, e.g., Khalil [58], but to help build more advanced stability concepts in the presence of deterministic

---

[1] To reduce notation, in the preface we assume that the initial condition x_0 is zero and make additional simplifications of statements; for a precise description of the results the reader is referred to the corresponding chapters.

and stochastic disturbances. Chapters 7 and 8 require the knowledge of invariant manifold theory, averaging, singular perturbations and some real analysis. Due to its length and a uniform presentation of robust/stochastic/adaptive nonlinear control, the book can be used for a second level course on nonlinear control.

Acknowledgments. The authors thank Petar Kokotovic who was the main source of feedback on the material during its research phase. We have also greatly benefited from interaction with Andrzej Banaszuk, Tamer Basar, Randy Freeman, Bill Helton, Clas Jacobson, Mrdjan Jankovic, Ioannis Kanellakopoulos, Zigang Pan, Laurent Praly, Rodolphe Sepulchre, Bob Skelton, Eduardo Sontag, Jing Sun, and Takis Tsiotras. We thank Zhonghua Li who helped with typing Chapters 2 and 5, and Hsin-hsiung Wang who helped with Chapter 8.

We gratefully acknowledge the support that we have received from National Science Foundation, Air Force Office of Scientific Research, United Technologies Research Center, and Ford Motor Company.

MIROSLAV KRSTIĆ
HUA DENG

San Diego, California, January 1998

Contents

I DETERMINISTIC SYSTEMS — 1

1 Stability and Regulation without Uncertainties — 3
- 1.1 Lyapunov and LaSalle Theorems — 3
- 1.2 Example: Strict-Feedback Systems and Backstepping — 6
- 1.3 Control Lyapunov Functions (clf) and Sontag's Formula — 7
- Notes and References — 10

2 Disturbance Attenuation — 13
- 2.1 Input-to-State Stability (ISS) — 13
- 2.2 Input-to-State Stabilizability and iss-clf's — 15
- 2.3 Inverse Optimal Gain Assignment — 17
- 2.4 Stability Margins — 26
- 2.5 Inverse Optimality via Backstepping — 32
- 2.6 Design for Strict-Feedback Systems — 34
- 2.7 Performance Estimates — 39
- Notes and References — 40

II STOCHASTIC SYSTEMS — 43

3 Stochastic Stability and Regulation — 45
- 3.1 Stochastic Lyapunov and LaSalle Like Theorems — 45
- 3.2 Stabilization in Probability via Backstepping — 50
- 3.3 Stochastic Control Lyapunov Functions — 57
- 3.4 Inverse Optimal Stabilization in Probability — 58
- 3.5 Inverse Optimal Backstepping — 62
- 3.6 Output-Feedback — 66
  - 3.6.1 Design for Output Feedback Systems — 66
  - 3.6.2 Inverse Optimal Output-Feedback Stabilization — 71
  - 3.6.3 Example — 73
- Notes and References — 74

4	**Stochastic Disturbance Attenuation**	**77**
	4.1 Noise-to-State Stability (NSS)	77
	4.2 Noise-to-State Stabilization and nss-clf's	79
	4.3 Design for Strict-Feedback Systems	82
	4.4 Inverse Optimal Noise-to-State Stabilization	84
	Notes and References	90

III ADAPTIVE CONTROL   93

5	**Deterministic Adaptive Tracking**	**95**
	5.1 The Adaptive Tracking Problem	95
	5.2 Stable Adaptive Tracking and atclf's	97
	5.3 Adaptive Backstepping	102
	5.4 Inverse Optimal Adaptive Tracking	105
	5.5 Inverse Optimality via Backstepping	111
	5.6 Design for Strict-Feedback Systems	114
	5.7 Transient Performance	119
	Notes and References	120
6	**Stochastic Adaptive Regulation**	**121**
	6.1 Adaptive Stochastic Backstepping Design	121
	6.2 Example	126
	Notes and References	127
7	**Invariant Manifolds and Asymptotic Properties**	**129**
	7.1 Closed-Loop Adaptive System	130
	7.2 Parameter Estimates Converge to Constant Values	132
	7.3 Decomposition of the Parameter Vector	136
	7.4 Center Manifold Analysis	138
	7.5 Categorization of Equilibria	144
	7.6 Invariant Manifolds	146
	7.7 Convergence to Stabilizing/Destabilizing Parameters	150
	7.7.1 Case $r = p$	151
	7.7.2 Case $r = 0$	151
	7.7.3 Case $0 < r < p$	156
	7.8 Disconnecting Adaptation	157
	Notes and References	161
8	**Extremum Seeking**	**163**
	8.1 Extremum Seeking—Problem Statement	163
	8.2 A Peak Seeking Scheme	165
	8.3 Averaging Analysis	167
	8.4 Singular Perturbation Analysis	170

Notes and References 172

Appendices 173

A Legendre-Fenchel Transform and Young's Inequality 175

B Measure of Global Invariant Manifolds 177

Bibliography 181

Index 191

Part I
DETERMINISTIC SYSTEMS

Chapter 1

Stability and Regulation without Uncertainties

This chapter is fully devoted to the basics of stability and stabilization theory for nonlinear systems *without uncertainties*. The concepts introduced here are then used in subsequent chapters for nonlinear *uncertain* systems. Some of the concepts that go beyond stabilization (like inverse optimality and associated margins)—which are treated in detail for systems with uncertainties in the subsequent chapters—are not studied in this chapter. For the inverse optimality theory for systems without uncertainties the reader is referred to [96]. The inverse optimality results from [96] can also obtained as a special case of the results in Chapter 2 by setting $g_1(x) \equiv 0$.

1.1 Lyapunov and LaSalle Theorems

Consider the nonautonomous system
$$\dot{x} = f(x,t) \tag{1.1}$$
where $f : \mathbb{R}^n \times \mathbb{R}_+ \to \mathbb{R}^n$ is locally Lipschitz in x and piecewise continuous in t.

Definition 1.1 *The origin $x = 0$ is the equilibrium point for (1.1) if*
$$f(0,t) = 0, \quad \forall t \geq 0. \tag{1.2}$$

Scalar *comparison functions* are important stability tools frequently used in this book.

Definition 1.2 *A continuous function $\gamma : \mathbb{R}_+ \to \mathbb{R}_+$ is said to belong to class $\mathcal{K}$ if it is strictly increasing and $\gamma(0) = 0$. It is said to belong to class $\mathcal{K}_\infty$ if $\gamma \in \mathcal{K}$ and $\gamma(r) \to \infty$ as $r \to \infty$.*

Definition 1.3 *A continuous function* $\beta : \mathbb{R}_+ \times \mathbb{R}_+ \to \mathbb{R}_+$ *is said to belong to class* $\mathcal{KL}$ *if for each fixed* s *the mapping* $\beta(r,s)$ *belongs to class* $\mathcal{K}_\infty$ *with respect to* r, *and for each fixed* r *the mapping* $\beta(r,s)$ *is decreasing with respect to* s *and* $\beta(r,s) \to 0$ *as* $s \to \infty$.

In this book we deal primarily with the problems of *global* stabilization of nonlinear systems. Next, we give definitions of equilibrium stability for deterministic systems.

Definition 1.4 *The equilibrium point* $x = 0$ *of (1.1) is*

- *globally uniformly stable, if there exists a class* $\mathcal{K}_\infty$ *function* $\gamma(\cdot)$ *such that*

$$|x(t)| \leq \gamma(|x(t_0)|), \quad \forall t \geq t_0 \geq 0, \quad \forall x(t_0) \in \mathbb{R}^n; \tag{1.3}$$

- *globally uniformly asymptotically stable, if there exists a class* $\mathcal{KL}$ *function* $\beta(\cdot,\cdot)$ *such that*

$$|x(t)| \leq \beta(|x(t_0)|, t - t_0), \quad \forall t \geq t_0 \geq 0, \quad \forall x(t_0) \in \mathbb{R}^n; \tag{1.4}$$

- *globally exponentially stable, if (1.4) is satisfied with* $\beta(r,s) = kre^{-\alpha s}$, $k > 0$, $\alpha > 0$.

The Lyapunov/LaSalle type theorems that follow are standard and can be found in any book on stability theory (for instance, in [58]). We give their statements and proofs for the convenience of the reader. The stochastic counterparts are given in section 3.1.

We start with a theorem that establishes global stability and certain convergence properties and then derive the main Lyapunov theorem on global asymptotic stability as a special case.

Theorem 1.5 (LaSalle-Yoshizawa) *Consider system (1.1) and suppose* $V : \mathbb{R}^n \times \mathbb{R}_+ \to \mathbb{R}_+$ *is a continuously differentiable function such that*

$$\gamma_1(|x|) \leq V(x,t) \leq \gamma_2(|x|) \tag{1.5}$$

$$\dot{V} = \frac{\partial V}{\partial t} + \frac{\partial V}{\partial x} f(x,t) \leq -W(x) \leq 0 \tag{1.6}$$

$\forall t \geq 0$, $\forall x \in \mathbb{R}^n$, *where* γ_1 *and* γ_2 *are class* $\mathcal{K}_\infty$ *functions and* W *is a continuous function. Then the equilibrium* $x = 0$ *is globally uniformly stable and*

$$\lim_{t \to \infty} W(x(t)) = 0. \tag{1.7}$$

1.1 Lyapunov and LaSalle Theorems

Proof. Since $\dot{V} \leq 0$, we have $\gamma_1(|x(t)|) \leq V(x(t),t) \leq V(x(t_0),t_0) \leq \gamma_2(|x(t_0)|)$, which means that $|x(t)| \leq \gamma_1^{-1}(\gamma_2(|x(t_0)|))$. This establishes global uniform stability and global uniform boundedness, that is, there exists a constant $B > 0$ such that $|x(t)| \leq B$, $\forall t \geq 0$. Since $V(x(t),t)$ is nonincreasing and bounded from below by zero, we conclude that it has a limit V_∞ as $t \to \infty$. Integrating (1.6), we have

$$\begin{aligned}
\lim_{t \to \infty} \int_{t_0}^{t} W(x(\tau))d\tau &\leq -\lim_{t \to \infty} \int_{t_0}^{t} \dot{V}(x(\tau),\tau)d\tau \\
&= \lim_{t \to \infty} \{V(x(t_0),t_0) - V(x(t),t)\} \\
&= V(x(t_0),t_0) - V_\infty,
\end{aligned} \quad (1.8)$$

which means that $\int_{t_0}^{\infty} W(x(\tau))d\tau$ exists and is finite. Now we show that $W(x(t))$ is also uniformly continuous. Since $|x(t)| \leq B$ and f is locally Lipschitz in x uniformly in t, we see that for any $t \geq t_0 \geq 0$,

$$\begin{aligned}
|x(t) - x(t_0)| &= \left|\int_{t_0}^{t} f(x(\tau),\tau)d\tau\right| \leq L\int_{t_0}^{t} |x(\tau)|d\tau \\
&\leq LB|t - t_0|,
\end{aligned} \quad (1.9)$$

where L is the Lipschitz constant of f on $\{|x| \leq B\}$. Choosing $\delta(\varepsilon) = \frac{\varepsilon}{LB}$, we have

$$|x(t) - x(t_0)| < \varepsilon, \quad \forall \, |t - t_0| \leq \delta(\varepsilon), \quad (1.10)$$

which means that $x(t)$ is uniformly continuous. Since W is continuous, it is uniformly continuous on the compact set $\{|x| \leq B\}$. From the uniform continuity of $W(x)$ and $x(t)$, we conclude that $W(x(t))$ is uniformly continuous. Hence, it satisfies the conditions of the following lemma:

Lemma 1.6 (Barbalat) *Consider the function $\phi : \mathbb{R}_+ \to \mathbb{R}$. If ϕ is uniformly continuous and $\lim_{t \to \infty} \int_0^\infty \phi(\tau)d\tau$ exists and is finite, then*

$$\lim_{t \to \infty} \phi(t) = 0. \quad (1.11)$$

Proof. Suppose that (1.11) does not hold; that is, either the limit does not exist or it is not equal to zero. Then there exists $\varepsilon > 0$ such that for every $T > 0$ one can find $t_1 \geq T$ with $|\phi(t_1)| > \varepsilon$. Since ϕ is uniformly continuous, there is a positive constant $\delta(\varepsilon)$ such that $|\phi(t) - \phi(t_1)| < \varepsilon/2$ for all $t_1 \geq 0$ and all t such that $|t - t_1| \leq \delta(\varepsilon)$. Hence, for all $t \in [t_1, t_1 + \delta(\varepsilon)]$, we have

$$\begin{aligned}
|\phi(t)| &= |\phi(t) - \phi(t_1) + \phi(t_1)| \\
&\geq |\phi(t_1)| - |\phi(t) - \phi(t_1)| \\
&> \varepsilon - \frac{\varepsilon}{2} = \frac{\varepsilon}{2},
\end{aligned} \quad (1.12)$$

which implies that

$$\left| \int_{t_1}^{t_1+\delta(\varepsilon)} \phi(\tau)d\tau \right| = \int_{t_1}^{t_1+\delta(\varepsilon)} |\phi(\tau)|d\tau > \frac{\varepsilon\delta(\varepsilon)}{2}, \tag{1.13}$$

where the first equality holds since $\phi(t)$ does not change sign on $[t_1, t_1 + \delta(\varepsilon)]$. Noting that $\int_0^{t_1+\delta(\varepsilon)} \phi(\tau)d\tau = \int_0^{t_1} \phi(\tau)d\tau + \int_{t_1}^{t_1+\delta(\varepsilon)} \phi(\tau)d\tau$, we conclude that $\int_0^t \phi(\tau)d\tau$ cannot converge to a finite limit as $t \to \infty$, which contradicts the assumption of the lemma. Thus, $\lim_{t\to\infty} \phi(t) = 0$.
□

Thus Lemma 1.6 guarantees that $W(x(t)) \to 0$ as $t \to \infty$. □

Theorem 1.7 (Main Lyapunov Theorem) *Consider system (1.1) and suppose $V : \mathbb{R}^n \times \mathbb{R}_+ \to \mathbb{R}_+$ is a continuously differentiable function such that*

$$\gamma_1(|x|) \leq V(x,t) \leq \gamma_2(|x|) \tag{1.14}$$

$$\dot{V} = \frac{\partial V}{\partial t} + \frac{\partial V}{\partial x}f(x,t) \leq -\gamma_3(|x|) \tag{1.15}$$

$\forall t \geq 0$, $\forall x \in \mathbb{R}^n$, *where γ_1 and γ_2 are class $\mathcal{K}_\infty$ functions and γ_3 is a class $\mathcal{K}$ function. Then the equilibrium $x = 0$ is globally uniformly asymptotically stable.*

Proof. This theorem is a direct corollary of Theorem 1.5 and the fact that $W(x) = \gamma_3(|x|)$ is positive definite. □

1.2 Example: Strict-Feedback Systems and Backstepping

The topic of this book is stabilization of nonlinear uncertain systems. In this section we introduce this problem for nonlinear systems *without uncertainties*. Consider the general class of systems affine in the control vector $u \in \mathbb{R}^m$:

$$\dot{x} = f(x) + g(x)u. \tag{1.16}$$

Not all the systems in this form are stabilizable. However, some interesting (and practically relevant) classes of nonlinear systems are stabilizable, such as, for example, the class of *strict-feedback systems*

$$\dot{x}_i = x_{i+1} + \varphi_i(\bar{x}_i), \quad i = 1,\ldots,n-1 \tag{1.17}$$
$$\dot{x}_n = u + \varphi(x), \tag{1.18}$$

where $\bar{x}_i = [x_1,\ldots,x_i]^T$, the functions $\varphi_i(\bar{x}_i)$ are smooth, and $\varphi_i(0) = 0$. This triangular class is only an example of a much broader class of systems for which globally stabilizing control laws are currently available [44, 63, 96].

1.3 CONTROL LYAPUNOV FUNCTIONS (CLF) AND SONTAG'S FORMULA

Theorem 1.8 *The control law*

$$z_i = x_i - \alpha_{i-1}(\bar{x}_{i-1}), \qquad \alpha_0 = 0 \qquad (1.19)$$

$$\alpha_i(\bar{x}_i) = -z_{i-1} - c_i z_i - \varphi_i + \sum_{j=1}^{i} \frac{\partial \alpha_{i-1}}{\partial x_j}(x_{j+1} + \varphi_j), \qquad c_i > 0 \qquad (1.20)$$

$$u = \alpha_n \qquad (1.21)$$

guarantees global asymptotic stability of the equilibrium $x = 0$ for the system (1.17), (1.18).

Proof. Consider the Lyapunov function

$$V = \frac{1}{2} \sum_{i=1}^{n} z_i^2, \qquad (1.22)$$

which is positive definite and radially unbounded in x because $\alpha_i(0) = 0$. It is straightforward to show that

$$\dot{V} = -\sum_{i=1}^{n} c_i z_i^2. \qquad (1.23)$$

This proves global asymptotic stability. $\square$

The control law given in Theorem 1.8 is a smooth function of x. Achieving smoothness, or at least continuity, in the measured state of the system, is important for practical implementation of a controller. Thus, throughout this book, we will require that the controllers designed be continuous.

1.3 Control Lyapunov Functions (clf) and Sontag's Formula

Let us start by considering the time derivative of a Lyapunov function candidate for the general system (1.16):

$$\begin{aligned} \dot{V} &= \frac{\partial V}{\partial x} f(x) + \frac{\partial V}{\partial x} g(x) u \\ &\triangleq L_f V + L_g V u. \end{aligned} \qquad (1.24)$$

The objective is to find a control law $u = \alpha(x)$ (hopefully continuous) that makes $\dot{V}$ a negative definite function of x. In general, the Lyapunov function is not given to the designer, so the search is both for a Lyapunov function and a control law. This brings us to the concept of a control Lyapunov function.

Definition 1.9 *A smooth positive definite radially unbounded function $V(x)$ is called a **control Lyapunov function (clf)** if*

$$\inf_{u \in \mathbb{R}^m} \{L_f V + L_g V u\} < 0, \qquad \forall x \neq 0. \tag{1.25}$$

An obvious alternative statement of this definition is given in the following lemma.

Lemma 1.10 *A smooth positive definite radially unbounded function V is a clf if and only if $L_g V = 0 \Rightarrow L_f V < 0$ whenever $x \neq 0$.*

The example in the previous section allows us to check this condition:

$$L_g V = z_n \tag{1.26}$$

$$L_f V = -\sum_{i=1}^{n-1} c_i z_i^2 + z_n \left[z_{n-1} + \varphi_n - \sum_{i=1}^{n} \frac{\partial \alpha_{n-1}}{\partial x_j} (x_{j+1} + \varphi_j) \right]. \tag{1.27}$$

Clearly, the condition $L_g V = 0 \Rightarrow L_f V < 0$ is satisfied.

The cornerstone of stability theory for systems without control is that the existence of a Lyapunov function implies stability. An analogous result exists for the stabilization problem and it states that the existence of a control Lyapunov function implies stabilizability. This result was first conceptually established for systems possibly non-affine in control by Artstein [3], and then constructively proved for systems affine in control by Sontag [99] who proposed a 'universal formula' for stabilization:

$$u = \alpha_s(x) = \begin{cases} -\dfrac{L_f V + \sqrt{(L_f V)^2 + (L_g V (L_g V)^T)^2}}{L_g V (L_g V)^T} (L_g V)^T, & L_g V \neq 0 \\ 0, & L_g V = 0 \end{cases} \tag{1.28}$$

This formula is globally asymptotically stabilizing because it gives

$$\dot{V} = -\sqrt{(L_f V)^2 + (L_g V (L_g V)^T)^2}, \tag{1.29}$$

which is positive definite by Lemma 1.10. The crucial question about $\alpha_s(x)$ is whether it is continuous. The first step towards answering this question is given by the following lemma.

Lemma 1.11 *The function*

$$\phi(a,b) = \begin{cases} 0 & \text{if } b = 0 \text{ and } a < 0 \\ -\dfrac{a + \sqrt{a^2 + b^2}}{b}, & \text{elsewhere} \end{cases} \tag{1.30}$$

is real analytic on the set $S = \{(a,b) \in \mathbb{R}^2 \mid b > 0 \text{ or } a < 0\}$.

1.3 CONTROL LYAPUNOV FUNCTIONS (CLF) AND SONTAG'S FORMULA

Proof. Consider the function

$$F(a, b, p) = bp^2 - 2ap - b, \tag{1.31}$$

which, after some calculations, gives $F(a, b, \phi(a, b)) = 0$, $\forall (a, b) \in S$. Another simple computation gives

$$\frac{\partial F(a, b, \phi(a, b))}{\partial p} = 2\sqrt{a^2 + b^2}, \tag{1.32}$$

which is nonzero on S. By the implicit function theorem, since F is real analytic, the solution $p = \phi(a, b)$ of $F(a, b, p) = 0$ is also real analytic. □

The next lemma establishes the smoothness of the Sontag formula away from the origin.

Lemma 1.12 *If $V(x)$ is a clf, then (1.28) is smooth away from $x = 0$.*

Proof. Since $V(x)$ is a clf, for each x the pair $(a, b) = (L_f V, L_g V (L_g V)^T)$ is in S. The control law (1.28) can be written as

$$\alpha_s(x) = \begin{cases} 0 & x = 0 \\ -\phi(L_f V, L_g V (L_g V)^T)(L_g V)^T, & x \neq 0 \end{cases} \tag{1.33}$$

As a composition of the real analytic function $\phi(a, b)$ and the smooth functions $L_f V(x)$ and $L_g V(x)$, this control law is smooth on $\mathbb{R}^n \setminus \{0\}$. □

The previous analysis did not establish continuity at the origin. To ensure it, we require the following property.

Definition 1.13 *A clf $V(x)$ is said to satisfy the* small control property (scp) *if there exists a control law $\alpha_c(x)$ continuous on $\mathbb{R}^n$ such that*

$$L_f V(x) + L_g V(x) \alpha_c(x) < 0, \quad \forall x \neq 0. \tag{1.34}$$

Lemma 1.14 *If $V(x)$ is a clf that satisfies the small control property, then (1.28) is continuous on $\mathbb{R}^n$.*

Proof. Because of Lemma 1.12, we only need to prove that $\alpha_s(x)$ is continuous at the origin. From (1.34) it follows that

$$|L_f V(x)| \leq |L_g V(x)||\alpha_c(x)|, \quad L_f V(x) \geq 0. \tag{1.35}$$

Then it follows that

$$|\alpha_s(x)| \leq |\alpha_c(x)| + \sqrt{|\alpha_c(x)|^2 + L_g V (L_g V)^T} \tag{1.36}$$

for $L_fV(x) \geq 0$. On the other hand, for $L_fV(x) < 0$, we have $0 \leq L_fV + \sqrt{(L_fV)^2 + (L_gV(L_gV)^{\mathrm{T}})^2} \leq L_gV(L_gV)^{\mathrm{T}}$, which implies that $|\alpha_s(x)| \leq |L_gV|$, and hence (1.36) holds for all x. Since $\alpha_c(x)$ and $L_gV(x)$ are both continuous, (1.36) implies that $\alpha_s(x)$ is continuous at the origin. □

Thus we have proved that the existence of a clf which satisfies the small control property implies stabilizability by a control law which is continuous at the origin and smooth everywhere else. The converse is immediate from a converse Lyapunov theorem. Hence, the following main result holds.

Theorem 1.15 (Sontag) *The system (1.16) is stabilizable by feedback continuous at the origin and smooth away from the origin if and only if there exists a clf with a small control property.*

In the subsequent chapters we will be applying Sontag-type formulae to various stabilization problems for nonlinear uncertain systems. The issue of continuity at the origin is always handled in the same way—by assuming the small control property. The *small control property* in a stabilization problem $\mathcal{X}$ is always defined as a property of the $\mathcal{X}$-control Lyapunov function that there exists a continuous control law $\alpha_c(x)$ which guarantees the form of stability $\mathcal{X}$ with respect to the $\mathcal{X}$-Lyapunov function $V(x)$. The proofs of continuity of various Sontag-type formulae always follow the reasoning in the proof of Lemma 1.14.

Notes and References

Stability is a classical subject with vast literature. Khalil [58] gives a comprehensive presentation of the theory and a review of the literature. Theorem 1.5 was proved by LaSalle [68] and Yoshizawa [116]. The technical lemma due to Barbalat appears in [91].

The control Lyapunov function approach to nonlinear stabilization was introduced by Artstein [3] and Sontag [98]. For control-affine systems, Sontag [99] also introduced the 'universal formula' presented in Section 1.3. The global stabilization theory for nonlinear systems also has its roots in the differential geometric control theory presented in the books of Isidori [44] and Nijmeier and Van der Schaft [85]. The state of the art in global stabilization of nonlinear systems with known models is well represented by the recent book of Sepulchre, Janković, and Kokotović [96].

The two main techniques used for construction of clf's and nonlinear controllers are *backstepping* (mentioned in Section 1.2) and *forwarding*. Backstepping is used extensively in this book and we will discuss the available literature in the Notes and References of subsequent chapters. Forwarding is not pursued in this book as it does not lend itself easily to the treatment of uncertainties. The three major sources on forwarding are the papers by

Teel [105] and Mazenc and Praly [78] and the book by Sepulchre, Janković, and Kokotović [96].

Chapter 2

Disturbance Attenuation

This chapter poses and solves the problem of disturbance attenuation for nonlinear systems affine in the disturbance. Two types of disturbance attenuation are studied: (1) input-to-state stabilization which relates the peak values of the disturbance and the system state, and (2) differential game problem which relates integral penalties on the disturbance and the state/controller pair. We show that these two problems are equivalent and present constructive analytical solutions. One way of seeing the results of the chapter is as a solution to a problem more general than the "nonlinear $\mathcal{H}_\infty$" problem without requiring that Hamilton-Jacobi-Isaacs partial differential equations be solved. The control approach presented in the chapter yields stability margins which are a nonlinear analog of Kalman's margins for the linear quadratic regulator (infinite gain margin, 60° phase margin).

2.1 Input-to-State Stability (ISS)

Consider the nonlinear system

$$\dot{x} = f(x,t) + g_1(x,t)d, \qquad (2.1)$$

where $x \in \mathbb{R}^n$ is the state, $d \in \mathbb{R}^r$ is the disturbance, and $f(0,t) \equiv 0$.

Definition 2.1 *The system (2.1) is said to be* input-to-state stable *(ISS) if there exist a class $\mathcal{KL}$ function β and a class $\mathcal{K}$ function χ, such that, for any $x(t_0)$ and for any input $d(\cdot)$ continuous on $[0,\infty)$ the solution exists for all $t \geq 0$ and satisfies*

$$|x(t)| \leq \beta(|x(t_0)|, t-t_0) + \chi\left(\sup_{t_0 \leq \tau \leq t} |d(\tau)|\right) \qquad (2.2)$$

for all t_0 and t such that $0 \leq t_0 \leq t$.

The following theorem establishes the connection between the existence of a Lyapunov-like function and the input-to-state stability.

Theorem 2.2 (Sontag) *Suppose that for the system (2.1) there exists a C^1 function $V : \mathbb{R}^n \times \mathbb{R}_+ \to \mathbb{R}_+$ such that for all $x \in \mathbb{R}^n$ and $d \in \mathbb{R}^m$,*

$$\gamma_1(|x|) \leq V(x,t) \leq \gamma_2(|x|) \tag{2.3}$$

$$|x| \geq \rho(|d|) \Rightarrow \frac{\partial V}{\partial t} + \frac{\partial V}{\partial x} f(x,t) + \frac{\partial V}{\partial x} g_1(x,t) d \leq -\gamma_3(|x|), \tag{2.4}$$

where γ_1, γ_2, and ρ are class $\mathcal{K}_\infty$ functions and γ_3 is a class $\mathcal{K}$ function. Then the system (2.1) is ISS with $\chi = \gamma_1^{-1} \circ \gamma_2 \circ \rho$.

Proof. If $x(t_0)$ is in the set

$$R_{t_0} = \left\{ x \in \mathbb{R}^n \,\middle|\, |x| \leq \rho\left(\sup_{\tau \geq t_0} |d(\tau)|\right) \right\}, \tag{2.5}$$

then $x(t)$ remains within the set

$$S_{t_0} = \left\{ x \in \mathbb{R}^n \,\middle|\, |x| \leq \gamma_1^{-1} \circ \gamma_2 \circ \rho\left(\sup_{\tau \geq t_0} |d(\tau)|\right) \right\} \tag{2.6}$$

for all $t \geq t_0$. Define $B = [t_0, T)$ as the time interval before $x(t)$ enters R_{t_0} for the first time. In view of the definition of R_{t_0}, we have

$$\dot{V} \leq -\gamma_3 \circ \gamma_2^{-1}(V), \qquad \forall t \in B. \tag{2.7}$$

Then, by [100, Lemma 6.1], there exists a class $\mathcal{KL}$ function β_V such that

$$V(t) \leq \beta_V(V(t_0), t - t_0), \qquad \forall t \in B, \tag{2.8}$$

which implies

$$|x(t)| \leq \gamma_1^{-1}(\beta_V(\gamma_2(|x(t_0)|), t - t_0)) \triangleq \beta(|x(t_0)|, t - t_0), \qquad \forall t \in B. \tag{2.9}$$

On the other hand, by (2.6), we conclude

$$|x(t)| \leq \gamma_1^{-1} \circ \gamma_2 \circ \rho\left(\sup_{\tau \geq t_0} |d(\tau)|\right) \triangleq \gamma\left(\sup_{\tau \geq t_0} |d(\tau)|\right), \qquad \forall t \in [t_0, \infty) \setminus B. \tag{2.10}$$

Then, by (2.9) and (2.10),

$$|x(t)| \leq \beta(|x(t_0)|, t - t_0) + \gamma\left(\sup_{\tau \geq t_0} |u(\tau)|\right), \qquad \forall t \geq t_0 \geq 0. \tag{2.11}$$

By causality, it follows that

$$|x(t)| \leq \beta(|x(t_0)|, t - t_0) + \gamma\left(\sup_{t_0 \leq \tau \leq t} |d(\tau)|\right), \qquad \forall t \geq t_0 \geq 0. \tag{2.12}$$

$\square$

A function V satisfying conditions of Theorem 2.2 is called an *ISS-Lyapunov function*. It was proved by Sontag and Wang [101] that the converse also holds, namely, that ISS implies the existence of an ISS-Lyapunov function.

2.2 Input-to-State Stabilizability and iss-clf's

Consider the system which, in addition to the disturbance input d, also has a control input u:
$$\dot{x} = f(x) + g_1(x)d + g_2(x)u, \qquad (2.13)$$
where $u \in \mathbb{R}^m$ and $f(0) = 0$. We have omitted the time dependence from f and g_1 for notational simplicity. We say that the system (2.13) is *input-to-state stabilizable* if there exists a control law $u = \alpha(x)$ continuous everywhere with $\alpha(0) = 0$, such that the closed-loop system is ISS with respect to d.

Definition 2.3 *A smooth positive definite radially unbounded function $V : \mathbb{R}^n \to \mathbb{R}_+$ is called an **ISS-control Lyapunov function (iss-clf)** for (2.13) if there exists a class $\mathcal{K}_\infty$ function ρ such that the following implication holds for all $x \neq 0$ and all $d \in \mathbb{R}^r$:*

$$|x| \geq \rho(|d|)$$
$$\Downarrow \qquad\qquad\qquad (2.14)$$
$$\inf_{u \in \mathbb{R}^m} \{L_f V + L_{g_1} V d + L_{g_2} V u\} < 0.$$

The following lemma is the key for our later use of the Sontag formula for input-to-state stabilization.

Lemma 2.4 *A pair (V, ρ), where $V(x)$ is positive definite and radially unbounded and $\rho \in \mathcal{K}_\infty$, satisfies (2.14) in Definition 2.3 if and only if*

$$L_{g_2} V(x) = 0 \;\Rightarrow\; L_f V(x) + |L_{g_1} V(x)| \rho^{-1}(|x|) < 0, \quad \forall x \neq 0. \qquad (2.15)$$

Proof. (Necessity) By Definition 2.3, if $x \neq 0$ and $L_{g_2} V = 0$ then
$$|x| \geq \rho(|d|)$$
$$\Downarrow \qquad\qquad\qquad (2.16)$$
$$L_f V + L_{g_1} V d < 0.$$

Now consider the particular input
$$d = \frac{(L_{g_1} V)^T}{|L_{g_1} V|} \rho^{-1}(|x|). \qquad (2.17)$$

This input satisfies the upper part of the implication (2.16):
$$\rho(|d|) = |x|. \qquad (2.18)$$

Therefore, substituting (2.17) into the lower part of (2.16), we conclude that, if $x \neq 0$ and $L_{g_2}V = 0$ then

$$L_f V + |L_{g_1} V| \rho^{-1}(|x|) < 0, \qquad (2.19)$$

that is, (2.15) is satisfied for $x \neq 0$.

(Sufficiency) For $|x| \geq \rho(|d|)$, using (2.15) we have

$$\begin{aligned}
&\inf_u \{L_f V + L_{g_1} V d + L_{g_2} V u\} \\
&\leq \inf_u \{L_f V + |L_{g_1} V| |d| + L_{g_2} V u\} \\
&\leq \inf_u \{L_f V + |L_{g_1} V| \rho^{-1}(|x|) + L_{g_2} V u\} \\
&< 0.
\end{aligned} \qquad (2.20)$$

This completes the proof. $\square$

Corollary 2.5 *The system (2.1) is ISS if and only if there exist a smooth positive definite radially unbounded function $V(x)$ and a class $\mathcal{K}_\infty$ function ρ such that*

$$L_f V(x) + |L_{g_1} V(x)| \rho^{-1}(|x|) < 0, \qquad \forall x \neq 0. \qquad (2.21)$$

The following theorem establishes equivalence between input-to-state stabilizability and the existence of an iss-clf. It extends Theorem 1.15 to systems affine in the disturbance.

Theorem 2.6 *The system (2.13) is input-to-state stabilizable if and only if there exists an iss-clf with small control property.*

Proof. The only if part is immediate by the converse theorem of Sontag and Wang [101]. To prove the if part, we define the following Sontag-type control law $u = \alpha_s(x)$,

$$\alpha_s = \begin{cases} -\dfrac{\omega + \sqrt{\omega^2 + (L_{g_2} V (L_{g_2} V)^T)^2}}{L_{g_2} V (L_{g_2} V)^T} (L_{g_2} V)^T, & (L_{g_2} V)^T \neq 0 \\ 0, & (L_{g_2} V)^T = 0, \end{cases} \qquad (2.22)$$

where

$$\omega = L_f V + |L_{g_1} V| \rho^{-1}(|x|). \qquad (2.23)$$

Before we verify that (2.22) is input-to-state stabilizing, we recall that the continuity of the control law away from the origin follows from Lemmas 1.11 and 2.4 and the continuity of ω and $L_{g_2} V$. The continuity at the origin follows

2.3 INVERSE OPTIMAL GAIN ASSIGNMENT

from the same arguments as in the proof of Lemma 1.14. To see that (2.22) is input-to-state stabilizing, we substitute it into (2.13) and get

$$\begin{aligned}\dot{V} &= L_f V + L_{g_1} V d - \omega - \sqrt{\omega^2 + (L_{g_2} V (L_{g_2} V)^{\mathrm{T}})^2} \\ &\leq -\sqrt{\omega^2 + (L_{g_2} V (L_{g_2} V)^{\mathrm{T}})^2} - |L_{g_1} V| \left[\rho^{-1}(|x|) - |d|\right].\end{aligned} \quad (2.24)$$

For $|x| \geq \rho(|d|)$ we have

$$\dot{V} \leq -\sqrt{\omega^2 + (L_{g_2} V (L_{g_2} V)^{\mathrm{T}})^2}, \quad (2.25)$$

which is negative definite due to Lemma 2.4. Then by Theorem 2.2, the closed-loop system is ISS. □

2.3 Inverse Optimal Gain Assignment

In this section we start the development of the *inverse* optimal approach, which is a prominent theme throughout this book. The subject of the section is disturbance attenuation, which we also refer to as gain assignment (with respect to the disturbance). This problem has traditionally been studied as a differential game problem, which has more recently been named "nonlinear $\mathcal{H}_\infty$". The solution to such a problem is equivalent to solving a Hamilton-Jacobi-Isaacs (HJI) partial differential equation.

To motivate the inverse (as opposed to the direct) approach, we start with a simple example where the HJI equation associated with the differential game problem is not only difficult to solve, but *impossible* to solve. Consider the scalar system

$$\dot{x} = u + x^2 d \quad (2.26)$$

and the differential game problem

$$\inf_u \sup_d \int_0^\infty \left(x^2 + u^2 - \gamma^2 d^2\right), \quad (2.27)$$

where $\gamma > 0$. The resulting HJI equation

$$\left(\frac{x^4}{\gamma^2} - 1\right)\left(\frac{\partial V}{\partial x}\right)^2 = -4x^2 \quad (2.28)$$

is not solvable outside of the interval $x \in (-\sqrt{\gamma}, \sqrt{\gamma})$, and the optimal control law

$$u^* = -\gamma \frac{x}{\sqrt{\gamma^2 - x^4}} \quad (2.29)$$

is not defined outside of this interval either. Contrary to the discouraging outcome of the *direct* problem, the *inverse* problem is solvable, and in this section we show more than one solution.

Definition 2.7 *The **inverse optimal gain assignment** problem for system (2.13) is solvable if there exist a class $\mathcal{K}_\infty$ function γ whose derivative γ' is also a class $\mathcal{K}_\infty$ function, a matrix-valued function $R_2(x)$ such that $R_2(x) = R_2(x)^\mathrm{T} > 0$ for all x, positive definite, radially unbounded functions $l(x)$ and $E(x)$, and a feedback law $u = \alpha(x)$ continuous everywhere with $\alpha(0) = 0$, which minimizes the cost functional*

$$J(u) = \sup_{d \in \mathcal{D}} \left\{ \lim_{t \to \infty} \left[E(x(t)) + \int_0^t \left(l(x) + u^\mathrm{T} R_2(x) u - \gamma(|d|) \right) d\tau \right] \right\}, \quad (2.30)$$

where $\mathcal{D}$ is the set of locally bounded functions of x.

The cost functional (2.30) puts penalty on the state and both the control and the disturbance. The state-dependent weight $R_2(x)$ on the control u is not allowed to vanish anywhere in $\mathbb{R}^n$ (and is, in fact, allowed to take infinite values in parts of the state space where the open-loop system is "well behaved" and zero control can be used). The penalty on the disturbance is allowed to be non-quadratic. (The purpose of the "terminal penalty" $E(x(t))$ is to avoid imposing an assumption that $x(t) \to 0$ as $t \to \infty$.) We point out that it is possible to allow a non-quadratic penalty on control too but that we do not pursue it to keep things notationally simpler. It turns out that a non-quadratic penalty on control is essential in the case of stochastic disturbances, so we allow a non-quadratic control penalty in the stochastic chapters of the book.

In the next theorem we provide a sufficient condition for the solvability of the inverse optimal gain assignment problem. This theorem is followed by a result in Theorem 2.10, which shows how to construct a control law that satisfies the condition in Theorem 2.8 for any nonlinear system that is input-to-state stabilizable.

Before we start our developments, let us introduce the following notation: For a class $\mathcal{K}_\infty$ function γ whose derivative exists and is also a class $\mathcal{K}_\infty$ function, $\ell\gamma$ denotes the transform

$$\ell\gamma(r) = r(\gamma')^{-1}(r) - \gamma\left((\gamma')^{-1}(r)\right), \quad (2.31)$$

where $(\gamma')^{-1}(r)$ stands for the inverse function of $\dfrac{d\gamma(r)}{dr}$. Using integration by parts, it is easy to show (Lemma A.1.a) that $\ell\gamma$ is equal to the Legendre-Fenchel transform

$$\ell\gamma(r) = \int_0^r (\gamma')^{-1}(s)\, ds, \quad (2.32)$$

which was brought into the control theory by Laurent Praly in [92].

2.3 INVERSE OPTIMAL GAIN ASSIGNMENT

Theorem 2.8 *Consider the auxiliary system of (2.13):*

$$\dot{x} = f(x) + g_1(x)\ell\gamma(2|L_{g_1}V|)\frac{(L_{g_1}V(x))^{\mathrm{T}}}{|L_{g_1}V|^2} + g_2(x)u \qquad (2.33)$$

where $V(x)$ is a Lyapunov function candidate and γ is a class $\mathcal{K}_\infty$ function whose derivative γ' is also a class $\mathcal{K}_\infty$ function. Suppose that there exists a matrix-valued function $R_2(x) = R_2(x)^{\mathrm{T}} > 0$ such that the control law

$$u = \alpha(x) = -R_2(x)^{-1}(L_{g_2}V)^{\mathrm{T}} \qquad (2.34)$$

globally asymptotically stabilizes (2.33) with respect to $V(x)$. Then the control law

$$u = \alpha^*(x) = \beta\alpha(x) = -\beta R_2^{-1}(L_{g_2}V)^{\mathrm{T}}, \qquad (2.35)$$

with any $\beta \geq 2$, solves the inverse optimal gain assignment problem for system (2.13) by minimizing the cost functional

$$J(u) = \sup_{d \in \mathcal{D}} \left\{ \lim_{t \to \infty} \left[2\beta V(x(t)) + \int_0^t \left(l(x) + u^{\mathrm{T}} R_2(x)u - \beta\lambda\gamma\left(\frac{|d|}{\lambda}\right) \right) d\tau \right] \right\} \qquad (2.36)$$

for any $\lambda \in (0, 2]$, where

$$\begin{aligned}
l(x) &= -2\beta\left[L_f V + \ell\gamma(2|L_{g_1}V|) - L_{g_2}VR_2^{-1}(L_{g_2}V)^{\mathrm{T}}\right] \\
&\quad + \beta(2-\lambda)\ell\gamma(2|L_{g_1}V|) + \beta(\beta-2)L_{g_2}VR_2^{-1}(L_{g_2}V)^{\mathrm{T}}.
\end{aligned} \qquad (2.37)$$

Proof. Since the control law (2.34) stabilizes the system (2.33), there exists a continuous positive definite function $W : \mathbb{R}^n \to \mathbb{R}_+$ such that

$$L_f V + \ell\gamma(2|L_{g_1}V|) - L_{g_2}VR_2^{-1}(L_{g_2}V)^{\mathrm{T}} \leq -W(x). \qquad (2.38)$$

We then have

$$l(x) \geq 2\beta W(x) + \beta(2-\lambda)\ell\gamma(2|L_{g_1}V|) + \beta(\beta-2)L_{g_2}VR_2^{-1}(L_{g_2}V)^{\mathrm{T}}. \qquad (2.39)$$

Since $\lambda \leq 2$, $\beta \geq 2$, $W(x)$ is positive definite, and $\ell\gamma$ is a class $\mathcal{K}_\infty$ function (Lemma A.1.c), we conclude that $l(x)$ is also positive definite. Therefore $J(u)$ defined in (2.36) is a meaningful cost functional that puts penalty on x, u and d. Substituting $l(x)$ into (2.36), it follows that

$$\begin{aligned}
J(u) &= \sup_{d \in \mathcal{D}} \left\{ \lim_{t \to \infty} \left[2\beta V(x(t)) + \int_0^t \left(-2\beta L_f V - \beta\lambda\ell\gamma(2|L_{g_1}V|) \right.\right.\right. \\
&\quad \left.\left.\left. + \beta^2 L_{g_2}VR_2^{-1}(L_{g_2}V)^{\mathrm{T}} + u^{\mathrm{T}}R_2 u - \beta\lambda\gamma\left(\frac{|d|}{\lambda}\right) \right) d\tau \right] \right\}
\end{aligned}$$

$$
\begin{aligned}
= &\ \sup_{d\in\mathcal{D}} \left\{ \lim_{t\to\infty} \left[2\beta V(x(t)) - 2\beta \int_0^t (L_f V + L_{g_1} V d + L_{g_2} V u)\, d\tau \right.\right.\\
&\ + \int_0^t \left(u^{\mathrm{T}} R_2 u + 2\beta L_{g_2} V u + \beta^2 L_{g_2} V R_2^{-1} (L_{g_2} V)^{\mathrm{T}} \right) d\tau \\
&\ \left.\left. - \int_0^t \left(\beta\lambda\gamma\left(\frac{|d|}{\lambda}\right) - 2\beta L_{g_1} V d + \beta\lambda\ell\gamma(2|L_{g_1}V|) \right) d\tau \right] \right\} \\
= &\ \sup_{d\in\mathcal{D}} \left\{ \lim_{t\to\infty} \left[2\beta V(x(t)) - 2\beta \int_0^t dV + \int_0^t (u-\alpha^*)^{\mathrm{T}} R_2(u-\alpha^*) d\tau \right.\right.\\
&\ -\beta \int_0^t \left[\lambda\gamma\left(\frac{|d|}{\lambda}\right) - \lambda\gamma\left((\gamma')^{-1}(2|L_{g_1}V|)\right) \right.\\
&\ \left.\left.\left. + 2\left(\lambda |L_{g_1}V|(\gamma')^{-1}(2|L_{g_1}V|) - L_{g_1}V d \right) \right] d\tau \right] \right\} \quad \text{(by (2.31))} \\
= &\ 2\beta V(x(0)) + \int_0^\infty (u-\alpha^*)^{\mathrm{T}} R_2 (u-\alpha^*) dt \\
&\ + \beta\lambda \sup_{d\in\mathcal{D}} \int_0^\infty \underbrace{\left[-\gamma\left(\frac{|d|}{\lambda}\right) + \gamma\left(\frac{|d^*|}{\lambda}\right) - \gamma'\left(\frac{|d^*|}{\lambda}\right) \frac{(d^*)^{\mathrm{T}}}{\lambda |d^*|}(d^* - d) \right]}_{\Pi(d,d^*)} dt,
\end{aligned}
$$
(2.40)

where

$$
d^*(x) = \lambda(\gamma')^{-1}(2|L_{g_1}V|) \frac{(L_{g_1}V)^{\mathrm{T}}}{|L_{g_1}V|}. \tag{2.41}
$$

By Lemma A.1.d, $\Pi(d, d^*)$ can be rewritten as

$$
\Pi(d, d^*) = -\gamma\left(\frac{|d|}{\lambda}\right) - \ell\gamma\left(\gamma'\left(\frac{|d^*|}{\lambda}\right)\right) + \gamma'\left(\frac{|d^*|}{\lambda}\right)\frac{(d^*)^{\mathrm{T}}}{|d^*|}\frac{d}{\lambda}. \tag{2.42}
$$

Then by Lemma A.2 we have

$$
\Pi(d,d^*) \leq -\gamma\left(\frac{|d|}{\lambda}\right) - \ell\gamma\left(\gamma'\left(\frac{|d^*|}{\lambda}\right)\right) + \gamma\left(\frac{|d|}{\lambda}\right) + \ell\gamma\left(\gamma'\left(\frac{|d^*|}{\lambda}\right)\right) = 0, \tag{2.43}
$$

and $\Pi(d, d^*) = 0$ if and only if $\dfrac{d}{\lambda} = (\gamma')^{-1}\left(\gamma'\left(\dfrac{|d^*|}{\lambda}\right)\right) \dfrac{d^*}{|d^*|}$, that is,

$$
\Pi(d, d^*) = 0 \quad \text{iff} \quad d = d^*. \tag{2.44}
$$

Thus

$$
\sup_{d\in\mathcal{D}} \int_0^\infty \Pi(d, d^*) dt = 0, \tag{2.45}
$$

and the "worst case" disturbance is given by (2.41). The minimum of (2.40) is reached with $u = \alpha^*$. Hence the control law (2.35) minimizes the cost functional (2.36). The value function of (2.30) is $J^*(x) = 2\beta V(x)$. □

2.3 INVERSE OPTIMAL GAIN ASSIGNMENT

The parameter $\beta \geq 2$ in the statement of Theorem 2.8 represents a design degree of freedom. The parameter λ (note that it parameterizes not only the penalty on the disturbance but also the penalty on the state, $l(x)$) indicates that the same control law is inverse optimal with respect to an entire family of different cost functionals.

Remark 2.9 Even though not explicit in the proof of Theorem 2.8, $V(x)$ solves the following family of *Hamilton-Jacobi-Isaacs* equations :

$$L_f V - \frac{\beta}{2} L_{g_2} V R_2(x)^{-1} (L_{g_2} V)^{\mathrm{T}} + \frac{\lambda}{2} \ell \gamma(2|L_{g_1} V|) + \frac{l(x)}{2\beta} = 0, \quad (2.46)$$

parameterized by $(\beta, \lambda) \in [2, \infty) \times (0, 2]$. It is easily seen from the proof of the above theorem that, for zero initial conditions, the achieved disturbance attenuation level is

$$\int_0^\infty \left[l(x) + u^{\mathrm{T}} R_2(x) u \right] dt \;\leq\; \beta \lambda \int_0^\infty \gamma\left(\frac{|d|}{\lambda} \right) dt. \quad (2.47)$$

□

In the next theorem we design controllers that are inverse optimal in the sense of Definition 2.7. We emphasize that these controllers are not restricted to disturbances with $\int_0^\infty \gamma(|d|) dt < \infty$ because they achieve input-to-state stability and allow any bounded (and persistent) d.

Theorem 2.10 *If the system (2.13) is input-to-state stabilizable, then the inverse optimal gain assignment problem is solvable.*

Proof. By the converse theorem of Sontag and Wang [101] (the only if part of Theorem 2.6), there exist an iss-clf $V(x)$ and a class $\mathcal{K}_\infty$ function ρ such that (2.14) is satisfied. We now show that there exist a class $\mathcal{K}_\infty$ function γ and a control law $u = \alpha(x)$ of the form (2.34) such that the auxiliary system (2.33) is stabilized. To this end, consider the following Sontag-type control law $u = \alpha_s(x)$ defined in (2.22). The continuity properties of this control law have been established in the proof of Theorem 2.6. We now show that the control law $u = \frac{1}{2}\alpha_s(x)$ is an input-to-state stabilizing controller for system (2.13). The derivative of V is

$$\begin{aligned}
\dot{V}|_{u=\frac{\alpha_s}{2}} &= L_f V + L_{g_1} V d - \frac{1}{2}\left(L_f V + |L_{g_1} V| \rho^{-1}(|x|)\right) \\
&\quad - \frac{1}{2}\sqrt{\omega^2 + (L_{g_2} V (L_{g_2} V)^{\mathrm{T}})^2} \\
&= \frac{1}{2}\underbrace{\left(L_f V + |L_{g_1} V|\rho^{-1}(|x|)\right)}_{\omega} - \frac{1}{2}\sqrt{\omega^2 + (L_{g_2} V (L_{g_2} V)^{\mathrm{T}})^2} \\
&\quad + L_{g_1} V d - |L_{g_1} V|\rho^{-1}(|x|) \\
&\leq -W(x) - |L_{g_1} V|\left[\rho^{-1}(|x|) - |d|\right], \quad (2.48)
\end{aligned}$$

where
$$W(x) = \frac{1}{2}\left[-\omega + \sqrt{\omega^2 + (L_{g_2}V(L_{g_2}V)^T)^2}\right], \qquad (2.49)$$
which is positive definite because $L_{g_2}V = 0 \Rightarrow \omega < 0$. Therefore, the control law $u = \frac{1}{2}\alpha_s(x)$ input-to-state stabilizes the system (2.13).

Next we show that there exists a class $\mathcal{K}_\infty$ function γ such that the control law $u = \frac{1}{2}\alpha_s(x)$ globally asymptotically stabilizes the auxiliary system (2.33) with respect to $V(x)$. From (2.48) it follows that
$$L_{f+g_2\frac{\alpha_s}{2}}V + |L_{g_1}V|\rho^{-1}(|x|) = -W(x). \qquad (2.50)$$
Since $|L_{g_1}V(x)|$ vanishes at the origin $x = 0$, there exists a class $\mathcal{K}_\infty$ function π such that
$$|L_{g_1}V| \leq \pi(|x|). \qquad (2.51)$$
Since $\rho^{-1} \circ \pi^{-1}$ is in class $\mathcal{K}_\infty$, there exists a class $\mathcal{K}_\infty$ function ζ whose derivative ζ' is also a class $\mathcal{K}_\infty$ function, such that
$$\zeta(2r) \leq r\rho^{-1}\left(\pi^{-1}(r)\right). \qquad (2.52)$$
Let us define
$$\gamma = \ell\zeta. \qquad (2.53)$$
From Lemma A.1.b it follows that $\ell\ell\zeta = \zeta$, which implies that
$$\ell\gamma(2r) \leq r\rho^{-1}\left(\pi^{-1}(r)\right). \qquad (2.54)$$
Then with (2.22) we have
$$\begin{aligned}\dot{V}\Big|_{(2.33)} &= L_{f+g_2\frac{\alpha_s}{2}}V + \ell\gamma(2|L_{g_1}V|) \\ &\leq L_{f+g_2\frac{\alpha_s}{2}}V + |L_{g_1}V|\rho^{-1} \circ \pi^{-1}(|L_{g_1}V|) \quad \text{(by (2.54))} \\ &\leq L_{f+g_2\frac{\alpha_s}{2}}V + |L_{g_1}V|\rho^{-1}(|x|) \quad \text{(by (2.51))} \\ &= -W(x), \quad \text{(by (2.50))}\end{aligned} \qquad (2.55)$$
which means that the system (2.33) is globally asymptotically stabilized.

Since the control law $\frac{1}{2}\alpha_s(x)$ is of the form (2.34) with $R_2(x) = R_2(x)^T > 0$ given by
$$R_2(x) = I \begin{cases} \dfrac{2L_{g_2}V(L_{g_2}V)^T}{\omega + \sqrt{\omega^2 + (L_{g_2}V(L_{g_2}V)^T)^2}}, & L_{g_2}V \neq 0 \\ \text{any positive real number}, & L_{g_2}V = 0, \end{cases} \qquad (2.56)$$

2.3 INVERSE OPTIMAL GAIN ASSIGNMENT

by Theorem 2.8, the control law $u = \alpha_s(x)$ is inverse optimal with respect to the cost functional (2.36) with the penalty on the state given by

$$\begin{aligned} l(x) &= 4\left[W(x) + |L_{g_1}V|\rho^{-1}(|x|) - \ell\gamma(2|L_{g_1}V|)\right] \\ &\geq 4W(x) = 2\left[-\omega + \sqrt{\omega^2 + (L_{g_2}V(L_{g_2}V)^T)^2}\right]. \end{aligned} \quad (2.57)$$

The function $l(x)$ is positive definite but not necessarily *radially unbounded*. We now modify the control law to achieve a new $\hat{l}(x)$ that is radially unbounded. Let us suppose that $W(x)$ is not radially unbounded. By following the procedure in [100, page 440], we can find a continuous function $\eta'(\cdot)$ such that

$$\eta'(r) \geq 1, \qquad \forall r \geq 0, \quad (2.58)$$

and

$$\eta'(V(x))W(x) \text{ is radially unbounded.} \quad (2.59)$$

Let us introduce a new iss-clf $\hat{V}(x) = \int_0^{V(x)} \eta'(s)ds$ (which is positive definite and radially unbounded) and apply the formula (2.22). The resulting penalty on control is

$$\hat{l}(x) \geq 4\hat{W}(x) = 2\left[-\hat{\omega} + \sqrt{\hat{\omega}^2 + (L_{g_2}\hat{V}(L_{g_2}\hat{V})^T)^2}\right], \quad (2.60)$$

where $L_{g_2}\hat{V} = \eta'(V)L_{g_2}V$ and $\hat{\omega} = L_f\hat{V} + |L_{g_1}\hat{V}|\rho^{-1}(|x|) = \eta'(V)\omega$. Thus (2.60) becomes

$$\begin{aligned} \hat{l}(x) &\geq 2\eta'(V)\left[-\omega + \sqrt{\omega^2 + (\eta'(V))^2(L_{g_2}V(L_{g_2}V)^T)^2}\right] \\ &= 4\eta'(V)W(x) + 2\eta'(V)\left[\sqrt{\omega^2 + (\eta'(V))^2(L_{g_2}V(L_{g_2}V)^T)^2}\right. \\ &\quad \left. - \sqrt{\omega^2 + (L_{g_2}V(L_{g_2}V)^T)^2}\right] \\ &\geq 4\eta'(V)W(x), \end{aligned} \quad (2.61)$$

which is radially unbounded. This completes the proof of Theorem 2.10. □

The following example illustrates Theorem 2.10.

Example 2.11 Consider the system

$$\dot{x} = u + x^2 d. \quad (2.62)$$

Since the system is scalar, we take $V = \frac{1}{2}x^2$ and get $L_{g_1}V = x^3$ and $L_{g_2}V = x$. Picking the class $\mathcal{K}_\infty$ function ρ as $\rho(r) = r$, we have $\omega = |L_{g_1}V|\rho^{-1}(|x|) = x^4$, and the control law based on (2.22) is

$$u = \alpha_s(x) = -\left(x^2 + \sqrt{x^4 + 1}\right)x = -\frac{2}{R_2}L_{g_2}V, \quad (2.63)$$

where
$$R_2(x) = \frac{2}{x^2 + \sqrt{x^4 + 1}} > 0, \quad \forall x. \tag{2.64}$$

Now let us choose the class $\mathcal{K}_\infty$ function π as $\pi(r) = r^3$, from (2.54) we can take $\ell\gamma(2r) = r^{4/3}$, and from (2.32) we get $\gamma'(r) = \frac{27}{4}r^3$ and $\gamma(r) = \frac{27}{16}r^4$. The control $u = \frac{\alpha_s}{2}$ is stabilizing for the auxiliary system (2.33) of (2.62), which has the form

$$\dot{x} = u + x^3, \tag{2.65}$$

because the derivative of the Lyapunov function along the solutions of (2.65) is

$$\dot{V}\big|_{u=\frac{\alpha_s}{2}} = -\frac{-x^2 + \sqrt{x^4 + 1}}{2}x^2. \tag{2.66}$$

By Theorem 2.8, with $\beta = \lambda = 2$, the control $u = \alpha_s(x)$ (2.63) is optimal with respect to the cost functional

$$J(u) = \sup_d \left\{ \lim_{t \to \infty} \left[2x(t)^2 + \int_0^t \left(\frac{2x^2}{x^2 + \sqrt{x^4 + 1}} + \frac{2u^2}{x^2 + \sqrt{x^4 + 1}} - \frac{27}{64}d^4 \right) d\tau \right] \right\} \tag{2.67}$$

with a value function $J^*(x) = 2x^2$.

If, instead, we choose the class $\mathcal{K}_\infty$ function ρ as $\rho(r) = r^{1/3}$, we will end up with a different controller as well as a different (quadratic with respect to d) cost functional. Now we have $\omega = |L_{g_1}V|\rho^{-1}(|x|) = x^6$ and the control law based on (2.22) becomes

$$u = \alpha_s(x) = -\left(x^4 + \sqrt{x^8 + 1}\right)x = -\frac{2}{R_2}L_{g_2}V, \tag{2.68}$$

where
$$R_2(x) = \frac{2}{x^4 + \sqrt{x^8 + 1}} > 0, \quad \forall x. \tag{2.69}$$

We keep the class $\mathcal{K}_\infty$ function π the same as before, from (2.54) we can take $\ell\gamma(2r) = r^2$, and from (2.32) we get $\gamma'(r) = 2r$ and $\gamma(r) = r^2$. The control $u = \frac{\alpha_s}{2}$ is stabilizing for the auxiliary system (2.33) of (2.62), which has the form

$$\dot{x} = u + x^5, \tag{2.70}$$

because the time derivative of $V(x)$ along the solutions of (2.70) is

$$\dot{V}\big|_{u=\frac{\alpha_s}{2}} = -\frac{-x^4 + \sqrt{x^8 + 1}}{2}x^2. \tag{2.71}$$

2.3 INVERSE OPTIMAL GAIN ASSIGNMENT

Theorem 2.8 with $\beta = \lambda = 2$ then tells us that the control $u = \alpha_s(x)$ is optimal with respect to the cost functional

$$J(u) = \sup_d \left\{ \lim_{t\to\infty} \left[2x(t)^2 + \int_0^t \left(\frac{2x^2}{x^4 + \sqrt{x^8+1}} + \frac{2u^2}{x^4 + \sqrt{x^8+1}} - d^2 \right) d\tau \right] \right\} \quad (2.72)$$

with a value function $J^*(x) = 2x^2$.

Unfortunately, neither in (2.67) nor in (2.72) is $l(x)$ radially unbounded (it is only positive definite). In the proof of Theorem 2.10 we remedy this by redesigning the iss-clf and applying the Sontag formula with the new iss-clf. Fortunately, for this scalar system, it is easy to go a step further and show that controller (2.68), written as $u = -2L_{g_2}\hat{V}$, where

$$\hat{V}(x) = \int_0^{V(x)} \frac{4r^2 + \sqrt{1+16r^4}}{2} dr \quad (2.73)$$

is optimal with respect to the cost functional

$$J(u) = \sup_d \left\{ \lim_{t\to\infty} \left[4\hat{V}(x(t)) + \int_0^t \left(x^2 + u^2 - \hat{\gamma}(|d|) \right) d\tau \right] \right\}, \quad (2.74)$$

where

$$\begin{aligned} \hat{\gamma}(r) &= 4\ell \left(\xi \circ \eta^{-1} \right)(r/2) \\ \xi(r) &= (r^4 + \sqrt{1+r^8}) r^6 / 2 \quad (2.75) \\ \eta(r) &= (r^4 + \sqrt{1+r^8}) r^3 . \end{aligned}$$

From these expressions it is easy to see that the penalty $\hat{\gamma}(|d|)$ is quadratic near $d = 0$ and $\mathcal{O}\left(|d|^{10/3}\right)$ as $|d| \to \infty$. A striking feature of the cost functional (2.74) is that it has unity weighting on control. In Section 2.4, we show that this can always be achieved for systems that are input-to-state stabilizable, and we derive *stability margins* associated with this property.

We stressed above that all of the controllers we derive guarantee ISS, namely, guarantee bounded solutions for bounded disturbances. For example, (2.68) guarantees that $|x(t)| \leq e^{-t}|x(0)| + \frac{1}{2}\|d\|_\infty$. □

Next, we show that input-to-state stabilizability is not only sufficient but also necessary for the solvability of the inverse optimal gain assignment problem.

Theorem 2.12 *If the inverse optimal gain assignment problem is solvable for the system (2.13), then (2.13) is input-to-state stabilizable.*

Proof. We only sketch the proof. If the inverse optimal gain assignment problem is solvable, then the following HJI equation is satisfied:

$$L_f V - L_{g_2} V R_2^{-1}(L_{g_2} V)^{\mathrm{T}} + \ell\gamma(2|L_{g_1} V|) = -\frac{1}{4}l(x). \tag{2.76}$$

Then, along the solutions of (2.13) with a control law $u = -R_2(x)^{-1}(L_{g_2}V)^{\mathrm{T}}$ we have

$$\dot{V} = L_f V - L_{g_2} V R_2^{-1}(L_{g_2} V)^{\mathrm{T}} + L_{g_1} V d. \tag{2.77}$$

By Lemma A.2, we get

$$\dot{V} \leq L_f V - L_{g_2} V R_2^{-1}(L_{g_2} V)^{\mathrm{T}} + \ell\gamma(2|L_{g_1} V|) + \gamma\left(\frac{|d|}{2}\right), \tag{2.78}$$

which with (2.76) results in

$$\dot{V} \leq -\frac{1}{4}l(x) + \gamma\left(\frac{|d|}{2}\right). \tag{2.79}$$

Since $l(x)$ is positive definite and radially unbounded, by Theorem 2.2 the system (2.13) with $u = -R_2(x)^{-1}(L_{g_2}V)^{\mathrm{T}}$ is ISS. $\square$

By combining Theorem 2.10 and Theorem 2.12, we get the following result.

Corollary 2.13 *The inverse optimal gain assignment problem for system (2.13) is solvable if and only if the system is input-to-state stabilizable.*

2.4 Stability Margins

The main benefit of inverse optimality is that the controller remains input-to-state stabilizing in the presence of a certain class of input uncertainties. In this section, we show that

1. To achieve these margins, it is sufficient to make $R_2(x) = I$.

2. $R_2(x) = I$ can be achieved for systems that are input-to-state stabilizable.

We first prove the latter statement and then characterize the margins.

Definition 2.14 (Small control property in the sense of Janković et al. [51][1] (scpj)) *An iss-clf $V(x)$ is said to be an* iss-clf-scpj *if there exists*

---
[1] The property (2.81) is actually weaker than the condition in Janković, Sepulchre and Kokotović [51] for $\mathrm{rank}\left\{\frac{\partial}{\partial x}(L_{g_2}V)^{\mathrm{T}}(0)\right\}$ to be equal to $\dim\{u\}$. However, this condition is verifiable for a given V, while (2.81) is not.

2.4 STABILITY MARGINS

a continuous control law $u = \alpha_c(x)$ such that, for all $x \neq 0$ and all $d \in \mathbb{R}^r$,

$$|x| \geq \rho(|d|)$$
$$\Downarrow \qquad (2.80)$$
$$L_f V + L_{g_1} V d + L_{g_2} V \alpha_c(x) < 0,$$

and, in addition,

$$\lim_{\varepsilon \to 0} \max_{|x|=\varepsilon} \frac{|\alpha_c(x)|}{|L_{g_2} V(x)|} < \infty. \qquad (2.81)$$

Without (2.81), this is the standard small control property.

Theorem 2.15 *If the system (2.13) has an iss-clf-scpj, then the inverse optimal gain assignment problem is solvable with $R_2(x) = I$.*

Proof. The proof extends ideas from [96, pp. 104–105] and [51]. From the proof of Theorem 2.10 we know $V(x)$ satisfies the Isaacs equation

$$L_f V - L_{g_2} V R_2(x)^{-1} (L_{g_2} V)^\mathrm{T} + |L_{g_1} V| \rho^{-1}(|x|) = -W(x), \qquad (2.82)$$

where $R_2(x)$ is defined in (2.56) and $W(x)$ is positive definite and radially unbounded. Our task in this proof is to show that there exist positive definite radially unbounded functions $\hat{V}(x)$ and $\hat{l}(x)$, and a class $\mathcal{K}_\infty$ function $\hat{\gamma}$ whose derivative $\hat{\gamma}'$ is also in class $\mathcal{K}_\infty$, such that the following Isaacs equation is satisfied

$$L_f \hat{V} - \frac{\beta}{2} |L_{g_2} \hat{V}|^2 + \frac{\lambda}{2} \ell \hat{\gamma}(2|L_{g_1} \hat{V}|) = -\frac{\hat{l}(x)}{2\beta}, \qquad (2.83)$$

in which case, according to Theorem 2.8, the control law

$$u = -\beta (L_{g_2} \hat{V})^\mathrm{T}, \qquad \beta \geq 2 \qquad (2.84)$$

solves the inverse optimal gain assignment problem with $\hat{R}_2(x) = I$. Since $V(x)$ is an iss-clf-scpj with some continuous $\alpha_c(x)$ such that

$$\underbrace{L_f V + |L_{g_1} V| \rho^{-1}(|x|)}_{\omega} + L_{g_2} V \alpha_c(x) < 0, \qquad \forall x \neq 0 \qquad (2.85)$$

it follows that

$$|\omega| \leq |L_{g_2} V| |\alpha_c(x)|, \qquad \text{for } \omega \geq 0. \qquad (2.86)$$

From (2.56) we get

$$R_2(x)^{-1} \leq \left(\frac{|\omega|}{|L_{g_2} V|^2} + \frac{1}{2} \right) \leq \left(\frac{|\alpha_c(x)|}{|L_{g_2} V(x)|} + \frac{1}{2} \right) I \qquad (2.87)$$

for $\omega \geq 0$. For $\omega < 0$ we have $0 \leq \omega + \sqrt{\omega^2 + (L_{g_2}V(L_{g_2}V)^T)^2} \leq L_{g_2}V(L_{g_2}V)^T$ which yields $R_2(x)^{-1} \leq 1/2I$, so (2.87) holds for all $x \in \mathbb{R}^n$. Lemmas 1.11 and 2.4, along with the continuity of $\omega(x)$ and $L_{g_2}V(x)$, imply that $R_2(x)^{-1}$ is continuous away from $x = 0$. This, along with (2.81) and (2.87), implies that there exists a continuous positive function $\varrho(V)$ such that

$$R_2(x)^{-1} \leq \varrho(V(x))I, \qquad \forall x \in \mathbb{R}^n. \tag{2.88}$$

(Such a function always exists since $V(x)$ is radially unbounded.) Consider

$$\hat{V}(x) = \int_0^{V(x)} \varrho(s)ds, \tag{2.89}$$

which is positive definite, radially unbounded (due to the positiveness of $\varrho(\cdot)$), and C^1. Multiplying (2.82) by $\varrho(V)$, we get

$$L_f\hat{V} - \frac{\beta}{2}|L_{g_2}\hat{V}|^2 + |L_{g_1}\hat{V}|\rho^{-1}(|x|)$$
$$= -\varrho(V)W(x) + L_{g_2}V\left(R_2^{-1} - \frac{\beta}{2}\varrho(V)I\right)(L_{g_2}V)^T\varrho(V)$$
$$\leq -\varrho(V)W(x). \tag{2.90}$$

Since $\varrho(V)L_{g_1}V(x)$ is continuous and vanishes at $x = 0$, there exists a class $\mathcal{K}_\infty$ function $\hat{\pi}$ such that $|\varrho(V)L_{g_1}V(x)| \leq \hat{\pi}(|x|)$. Similar to the proof of Theorem 2.10, let $\hat{\gamma} = \ell\hat{\zeta}$, where $\hat{\zeta}$ is a class $\mathcal{K}_\infty$ function with a class $\mathcal{K}_\infty$ derivative selected so that $\hat{\zeta}(2r) \leq r\rho^{-1}(\hat{\pi}^{-1}(r))$. Then

$$\ell\hat{\gamma}(2|L_{g_1}\hat{V}|) \leq |L_{g_1}\hat{V}|\rho^{-1}\left(\hat{\pi}^{-1}(|L_{g_1}\hat{V}|)\right) \leq |L_{g_1}\hat{V}|\rho^{-1}(|x|). \tag{2.91}$$

Substituting this into (2.90) yields

$$L_f\hat{V} - \frac{\beta}{2}|L_{g_2}\hat{V}|^2 + \frac{\lambda}{2}\ell\hat{\gamma}(2|L_{g_1}\hat{V}|)$$
$$\leq -\varrho(V)W(x) + \frac{\lambda}{2}\ell\hat{\gamma}(2|L_{g_1}\hat{V}|) - |L_{g_1}\hat{V}|\rho^{-1}(|x|)$$
$$\leq -\varrho(V)W(x). \tag{2.92}$$

Thus $\hat{V}(x)$ satisfies the Isaacs equation (2.83) with $\hat{l}(x) \geq 2\beta\varrho(V)W(x)$, which is positive definite and radially unbounded. $\square$

Next, we derive the stability margins. In order to characterize the class of allowable input uncertainties, we remind the reader of the definition of *strict passivity* [11].

Definition 2.16 *The system*

$$\begin{aligned}\dot{\chi} &= \check{f}(\chi) + \check{g}(\chi)\check{u} \\ \check{y} &= \check{h}(\chi)\end{aligned} \tag{2.93}$$

2.4 STABILITY MARGINS

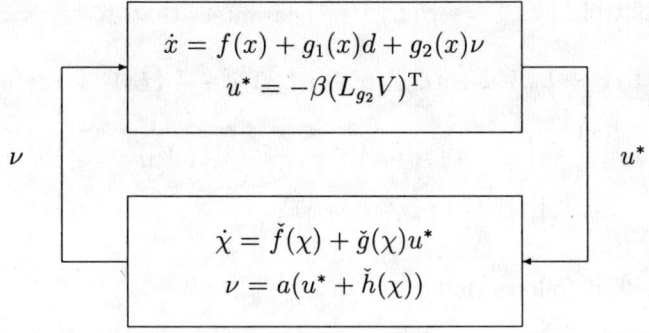

Figure 2.1: The composite system (2.95) is ISS with respect to d.

is said to be strictly passive if there exists a $\mathcal{C}^1$ positive definite radially unbounded (storage) function $\check{V}(\chi)$ and a class $\mathcal{K}_\infty$ (dissipation rate) function $\psi(\cdot)$ such that

$$\int_0^t \check{y}^T \check{u} d\sigma \geq \check{V}(\chi(t)) - \check{V}(\chi(0)) + \int_0^t \psi(|\chi(\sigma)|)d\sigma, \qquad (2.94)$$

for all $\check{u} \in \mathcal{C}^0$, $\chi(0) \in \mathbb{R}^{\check{n}}$, $t \geq 0$.

Theorem 2.17 *If a controller solves the inverse optimal gain assignment problem for (2.13) with $R_2(x) = I$, then it is input-to-state stabilizing for the system*

$$\begin{aligned} \dot{x} &= f(x) + g_1(x)d + g_2(x)a(u + \check{y}) \\ \dot{\chi} &= \check{f}(\chi) + \check{g}(\chi)u, \qquad \check{y} = \check{h}(\chi), \end{aligned} \qquad (2.95)$$

where $a \in [1/2, \infty)$ and the χ-system is strictly passive.

In simple words, an inverse optimal ISS controller remains ISS stabilizing through unmodeled dynamics of the form $a(I + \mathcal{P})$ where $\mathcal{P}$ is strictly passive, as depicted in Figure 2.1.

Proof. From the assumptions of the theorem we know that there exist Lyapunov-type functions $V(x)$ and $\check{V}(\chi)$ such that

$$L_f V - \frac{\beta}{2}|L_{g_2}V|^2 = -\frac{l(x)}{2\beta} - \frac{\lambda}{2}\ell\gamma(2|L_{g_1}V|) \qquad (2.96)$$

$$L_{\check{f}}\check{V} + L_{\check{g}}\check{V}u \leq -\psi(|\chi|) + \check{y}^T u, \qquad (2.97)$$

with l, γ, and ψ as in Definitions 2.7 and 2.16. Consider the following candidate for a composite iss-clf:

$$V_c(x,\chi) = V(x) + \frac{a}{\beta}\check{V}(\chi). \qquad (2.98)$$

Then the control law $u^* = -\beta(L_{g_2}V)^T$ guarantees that

$$\begin{aligned}
\dot{V}_c &= L_f V + L_{g_1}V d + a L_{g_2}V u^* + a L_{g_2}V \check{y} + \frac{a}{\beta}\left(L_{\check{f}}\check{V} + L_{\check{g}}\check{V}u^*\right) \\
&\leq -\frac{l(x)}{2\beta} + \left(\frac{1}{2} - a\right)\beta|L_{g_2}V|^2 - \frac{\lambda}{2}\ell\gamma(2|L_{g_1}V|) + L_{g_1}V d + a L_{g_2}V \check{y} \\
&\quad - \frac{a}{\beta}\psi(|\chi|) + \frac{a}{\beta}\check{y}^T\left(-\beta(L_{g_2}V)^T\right).
\end{aligned} \quad (2.99)$$

Since $a \geq 1/2$, it follows that

$$\dot{V}_c \leq -\frac{l(x)}{2\beta} - \frac{a}{\beta}\psi(|\chi|) - \frac{\lambda}{2}\ell\gamma(2|L_{g_1}V|) + \frac{\lambda}{2}2|L_{g_1}V|\frac{|d|}{\lambda}. \quad (2.100)$$

By applying Lemma A.2 to the last term, we get

$$\dot{V}_c \leq -\frac{l(x)}{2\beta} - \frac{a}{\beta}\psi(|\chi|) + \gamma\left(\frac{|d|}{\lambda}\right). \quad (2.101)$$

Since $l(x)$ and $\psi(|\chi|)$ are both radially unbounded, by Theorem 2.2, the closed-loop system is ISS. □

Theorems 2.15 and 2.17 can be combined to obtain the following corollary.

Corollary 2.18 *If the system (2.13) has an iss-clf-scpj, then there exists a control law that achieves input-to-state stability in the presence of input unmodeled dynamics of the form $a(I + \mathcal{P})$ with $a \geq 1/2$ and $\mathcal{P}$ strictly passive.*

By setting $d = 0$, we recover the result in [96]. In the linear case, this result implies the standard result that inverse optimal controllers possess infinite gain margins and 60° phase margins [1].

Example 2.19 (Example 2.1, continued) Consider the system (2.62). In Section 2.3, it was shown that the control law (2.68) is optimal with respect to the cost functional (2.72). By choosing

$$\varrho(r) = \frac{4r^2 + \sqrt{1 + 16r^4}}{2}, \quad (2.102)$$

it is easy to show that (2.88) is satisfied. Consider

$$\hat{V}(x) = \int_0^{V(x) = \frac{1}{2}x^2} \frac{4r^2 + \sqrt{1 + 16r^4}}{2} dr, \quad (2.103)$$

which is the same as (2.73). The control law

$$u = -2L_{g_2}\hat{V} = -\left(x^4 + \sqrt{x^8 + 1}\right)x \quad (2.104)$$

2.4 STABILITY MARGINS

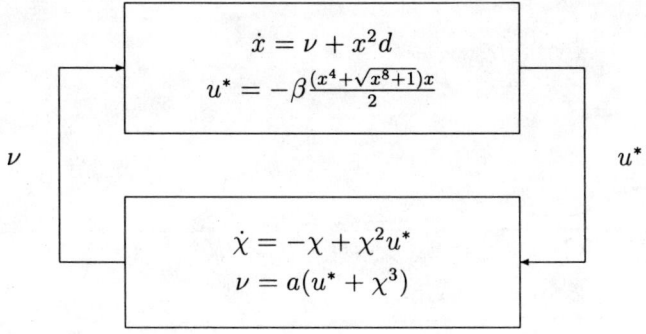

Figure 2.2: The composite system is ISS with respect to d.

is optimal with respect to the same cost functional (2.74) (for $\beta = \lambda = 2$). In summary, we have been able to set $R_2(x) = 1$ and to make $l(x)$ radially unbounded.

Let us now consider the strictly passive perturbation system

$$\begin{aligned} \dot{\chi} &= -\chi + \chi^2 \check{u} \\ \check{y} &= \chi^3, \end{aligned} \quad (2.105)$$

The nonlinear operator for $\mathcal{P}: \check{u} \mapsto \check{y}$ is not bounded (let alone small in $\mathcal{H}_\infty$ sense). It is not even ISS. But it does not retard the phase by more than 90°. By Theorem 2.17, the system shown in Figure 2.2 is ISS. For a better quantitative understanding, we repeat calculations that show ISS:

$$\begin{aligned} \frac{d}{dt}\left(\hat{V}(x) + \frac{a\chi^2}{2\beta}\right) &= \frac{\partial \hat{V}}{\partial x}(au^* + a\chi^3 + x^2 d) + \frac{2a}{\beta}\chi(-\chi + \chi^2 u^*) \\ &= \frac{x^4 + \sqrt{x^8+1}}{2}x\left[-\frac{a\beta}{2}(x^4 + \sqrt{x^8+1})x + a\chi^3 + x^2 d\right] \\ &\quad + \frac{a}{\beta}\chi\left[-\chi - \frac{\beta}{2}\chi^2(x^4 + \sqrt{x^8+1})x\right] \\ &= -\frac{a\beta}{4}x^2 - \frac{a}{\beta}\chi^2 - a\beta\frac{x^4 + \sqrt{x^8+1}}{2}x^6 \\ &\quad + \frac{x^4 + \sqrt{x^8+1}}{2}x^3 d \\ &\leq -\frac{a\beta}{4}x^2 - \frac{a}{\beta}\chi^2 + \hat{\gamma}\left(\frac{|d|}{2}\right) \quad (2.106) \end{aligned}$$

where the last inequality is obtained by applying Lemma A.2, using (2.75) and by noting that $a\beta \geq 1$. From (2.106) it follows that the system in Figure 2 is ISS with respect to d.

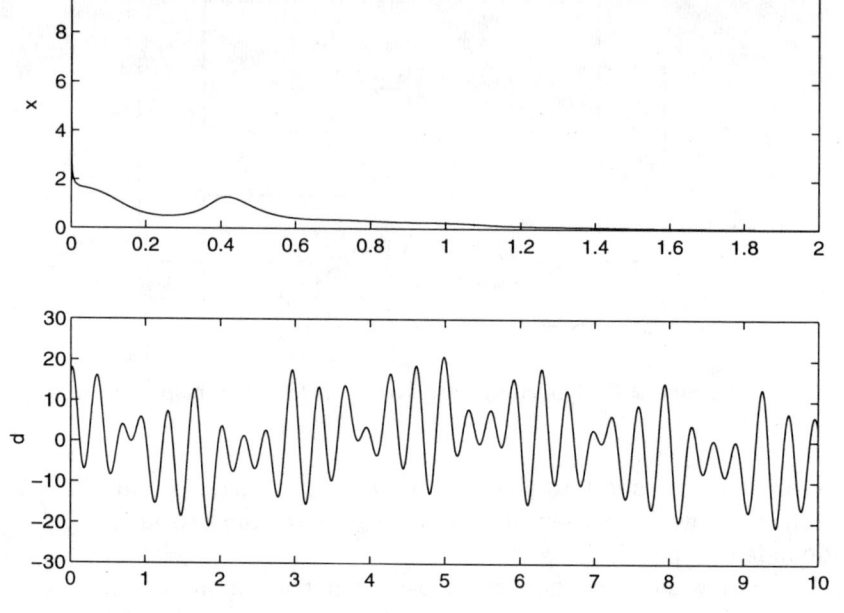

Figure 2.3: Trajectories of state x and disturbance d

Figure 2.3 shows the simulation results for an almost periodic disturbance $d(t) = 3\sin 9t + 5\sin\sqrt{13}t + 7\cos 15t + 9\cos 19t$, $\beta = \lambda = 2$, $a = 1/2$, and initial conditions $x(0) = \chi(0) = 10$. □

2.5 Inverse Optimality via Backstepping

In Section 2.3 we showed that the inverse optimal gain assignment problem reduces to the problem of finding an iss-clf. In this section, we show that integrator backstepping can be used for systematically constructing iss-clf's.

Lemma 2.20 *If the system*

$$\dot{x} = f(x) + g_1(x)d + g_2(x)u \qquad (2.107)$$

is input-to-state stabilizable with a smooth control law $u = \alpha(x)$, *then the augmented system*

$$\begin{aligned}\dot{x} &= f(x) + g_1(x)d + g_2(x)\xi \\ \dot{\xi} &= u\end{aligned} \qquad (2.108)$$

is also input-to-state stabilizable with a smooth control law.

Proof. It was proven by Sontag and Wang [101] that a system is ISS if and only if there exist a smooth positive definite radially unbounded function $V(x)$

2.5 INVERSE OPTIMALITY VIA BACKSTEPPING

and class $\mathcal{K}_\infty$ functions μ and ν such that the following 'dissipation' inequality holds

$$L_{f+g_2\alpha}V + L_{g_1}Vd \leq -\mu(|x|) + \nu(|d|). \tag{2.109}$$

We now show that the control law

$$u = \bar{\alpha}(x,\xi) = -L_{g_2}V - (\xi - \alpha) + L_{f+g_2\xi}\alpha - |L_{g_1}\alpha|^2(\xi - \alpha) \tag{2.110}$$

achieves input-to-state stabilization of (2.108) with respect to an iss-clf

$$\bar{V}(x,\xi) = V(x) + \frac{1}{2}(\xi - \alpha(x))^2. \tag{2.111}$$

Towards this end, consider

$$\begin{aligned}\dot{\bar{V}} &= L_{f+g_2\alpha}V + L_{g_1}Vd \\ &+ (\xi - \alpha)\left[u + L_{g_2}V - L_{f+g_2\xi}\alpha - (L_{g_1}\alpha)d\right].\end{aligned} \tag{2.112}$$

By substituting (2.109) and (2.110) we get

$$\begin{aligned}\dot{\bar{V}} &\leq -\mu(|x|) + \nu(|d|) - (\xi - \alpha)^2 - |L_{g_1}\alpha|^2(\xi - \alpha)^2 - (\xi - \alpha)(L_{g_1}\alpha)d \\ &\leq -\mu(|x|) + \nu(|d|) - (\xi - \alpha)^2 + \frac{1}{4}|d|^2.\end{aligned} \tag{2.113}$$

Denoting $\bar{\nu}(r) = \nu(r) + \frac{1}{4}r^2$ and picking a class $\mathcal{K}_\infty$ function $\bar{\mu}(r) \leq \min\{\mu(r), r^2\}$, we get

$$\begin{aligned}\dot{\bar{V}} &\leq -\bar{\mu}\left(\left|\begin{bmatrix}x \\ \xi - \alpha(x)\end{bmatrix}\right|\right) + \bar{\nu}(|d|) \\ &\leq -\tilde{\mu}\left(\left|\begin{bmatrix}x \\ \xi\end{bmatrix}\right|\right) + \bar{\nu}(|d|),\end{aligned} \tag{2.114}$$

where $\tilde{\mu} \in \mathcal{K}_\infty$. Thus, the control law (2.110) achieves input-to-state stabilization of (2.108). To see that $\bar{V}(x)$ is an iss-clf, we choose $\rho = \tilde{\mu}^{-1} \circ 2\bar{\nu}$. □

Theorem 2.21 *Under the conditions of Lemma 2.20, the inverse optimal gain assignment problem is solvable for system (2.108).*

Proof. Immediate, by combining Lemma 2.20 with Theorem 2.10. The continuity at the origin follows from the fact that in Lemma 2.20 we found a smooth input-to-state stabilizing control law, which implies that $\bar{V}(x,\xi)$ satisfies a small control property, and therefore the Sontag-type controller in Theorem 2.10 is continuous at the origin. □

A recursive application of Lemma 2.20, combined with Theorem 2.10 leads to the following result for a representative class of strict-feedback systems.

Corollary 2.22 *The inverse optimal gain assignment problem is solvable for the following system:*

$$\begin{aligned}\dot{x}_i &= x_{i+1} + \varphi_i(x_1,\ldots,x_i)^{\mathrm{T}}d, \qquad i=1,\ldots,n-1 \\ \dot{x}_n &= u + \varphi_n(x_1,\ldots,x_n)^{\mathrm{T}}d.\end{aligned} \qquad (2.115)$$

2.6 Design for Strict-Feedback Systems

Since the control law for strict-feedback systems (2.115) suggested by Corollary 2.22 is based on a Sontag-type formula, and is typically non-smooth (especially at the origin), in this section we design inverse optimal control laws that are *smooth everywhere*. In addition, they achieve a quadratic penalty on the disturbance with a constant weight function.

From Theorem 2.8, it follows that in order to solve the inverse optimal gain assignment problem, it suffices to find a stabilizing controller of the form (2.34) for the auxiliary system (2.33). We choose

$$\gamma(r) = \frac{1}{\kappa} r^2. \qquad (2.116)$$

With this choice, the auxiliary system (2.33) takes the form

$$\dot{x} = \begin{bmatrix} x_2 \\ x_3 \\ \vdots \\ x_n \\ 0 \end{bmatrix} + \kappa g_1 (L_{g_1} V)^{\mathrm{T}} + \begin{bmatrix} 0 \\ 0 \\ \vdots \\ 0 \\ 1 \end{bmatrix} u, \qquad (2.117)$$

where $g_1 = [\varphi_1, \varphi_2, \cdots, \varphi_n]^{\mathrm{T}}$.

First, we search for an iss-clf for the system (2.115). Repeated application of Lemma 2.20 gives an iss-clf

$$\begin{aligned} V &= \frac{1}{2}\sum_{i=1}^{n} z_i^2 \\ z_i &= x_i - \alpha_{i-1}(x_1,\cdots,x_i), \quad i=1,\cdots,n \end{aligned} \qquad (2.118)$$

where α_i's are to be determined. For notational convenience we define $z_0 := 0$, $\alpha_0 := 0$, $x_{n+1} = 0$ and $z_{n+1} = 0$. We then have

$$(L_{g_1} V)^{\mathrm{T}} = \sum_{j=1}^{n} \frac{\partial V}{\partial x_j} \varphi_j = \sum_{j=1}^{n} \left(z_j - \sum_{k=j+1}^{n} \frac{\partial \alpha_{k-1}}{\partial x_j} z_k \right) \varphi_j = \sum_{j=1}^{n} w_j z_j \quad (2.119)$$

$$\frac{\partial V}{\partial x} \begin{bmatrix} x_2 \\ x_3 \\ \vdots \\ x_n \\ 0 \end{bmatrix} = \sum_{j=1}^{n} \frac{\partial V}{\partial x_j} x_{j+1} = \sum_{j=1}^{n} \left(x_{j+1} - \sum_{k=1}^{j-1} \frac{\partial \alpha_{j-1}}{\partial x_k} x_{k+1} \right) z_j, \quad (2.120)$$

2.6 DESIGN FOR STRICT-FEEDBACK SYSTEMS

where

$$w_j(\bar{x}_j) = \varphi_j - \sum_{k=1}^{j-1} \frac{\partial \alpha_{j-1}}{\partial x_k} \varphi_k . \tag{2.121}$$

The functions $\alpha_1, \cdots, \alpha_{n-1}$ are sought to make V defined in (2.118) a clf for system (2.117). The derivative of V along the solutions of (2.117) is

$$\begin{aligned}\dot{V} &= z_n u + \kappa \left(\sum_{k=1}^{n} w_k z_k\right)^{\mathrm{T}} \left(\sum_{i=1}^{n} w_i z_i\right) + \sum_{i=1}^{n}\left(x_{i+1} - \sum_{k=1}^{i-1} \frac{\partial \alpha_{i-1}}{\partial x_k} x_{k+1}\right) z_i \\ &= z_n u + \kappa |w_n|^2 z_n^2 + \kappa z_n \left(\sum_{k=1}^{n-1} w_k^{\mathrm{T}} z_k\right) w_n \\ &\quad - z_n \sum_{k=1}^{n-1} \frac{\partial \alpha_{n-1}}{\partial x_k} x_{k+1} + \kappa \sum_{i=1}^{n-1}\left(\sum_{k=i+1}^{n} w_k^{\mathrm{T}} z_k\right) w_i z_i + z_n z_{n-1} \\ &\quad + \sum_{i=1}^{n-1}\left[z_{i-1} + \alpha_i + \kappa \left(\sum_{k=1}^{i} w_k^{\mathrm{T}} z_k\right) w_i - \sum_{k=1}^{i-1} \frac{\partial \alpha_{i-1}}{\partial x_k} x_{k+1}\right] z_i . \end{aligned} \tag{2.122}$$

The choice

$$\alpha_i = -z_{i-1} - c_i z_i - \kappa |w_i|^2 z_i - 2\kappa \left(\sum_{k=1}^{i-1} w_k^{\mathrm{T}} z_k\right) w_i + \sum_{k=1}^{i-1} \frac{\partial \alpha_{i-1}}{\partial x_k} x_{k+1} , \tag{2.123}$$

where $c_i > 0$, results in

$$\begin{aligned}\dot{V} &= -\sum_{k=1}^{n-1} c_k z_k^2 + z_n \left[u + z_{n-1} + \kappa |w_n|^2 z_n \right. \\ &\quad \left. + 2\kappa \left(\sum_{k=1}^{n-1} w_k^{\mathrm{T}} z_k\right) w_n - \sum_{k=1}^{n-1} \frac{\partial \alpha_{n-1}}{\partial x_k} x_{k+1}\right]. \end{aligned} \tag{2.124}$$

In the derivation of (2.124), we have used the equality

$$\kappa \sum_{i=1}^{n-1}\left(\sum_{k=i+1}^{n} w_k^{\mathrm{T}} z_k\right) w_i z_i - \kappa \sum_{i=1}^{n-1}\left(\sum_{k=1}^{i-1} w_k^{\mathrm{T}} z_k\right) w_i z_i = \kappa z_n \left(\sum_{k=1}^{n-1} w_k^{\mathrm{T}} z_k\right) w_n . \tag{2.125}$$

We are now in the position to choose the control u. We may choose u such that all the terms inside the bracket in (2.124) are cancelled and the bracketed term multiplying z_n is equal to $-c_n z_n^2$ as in [63], but the controller designed in that way is not guaranteed to be inverse optimal. In order for a controller to be inverse optimal, according to Theorem 2.8, it should be of the form

$$u = \alpha_n(x) = -R_2(x)^{-1}(L_{g_2} V)^{\mathrm{T}} , \tag{2.126}$$

where $R_2(x) = R_2(x)^{\mathrm{T}} > 0$, for all x. In light of (2.115) and (2.118), (2.126) simplifies to

$$u = \alpha_n(x) = -R_2(x)^{-1} z_n , \tag{2.127}$$

i.e., we must choose α_n with z_n as a factor.

Since $x_{k+1} = z_{k+1} + \alpha_k$, $k = 1, \cdots, n-1$, and each α_k vanishes at $z = 0$, there exist smooth functions ϕ_k, $k = 1, \cdots, n$, such that

$$-\sum_{k=1}^{n-1} \frac{\partial \alpha_{n-1}}{\partial x_k} x_{k+1} = \sum_{k=1}^{n} \phi_k z_k. \tag{2.128}$$

Thus (2.124) becomes

$$\dot{V} = -\sum_{k=1}^{n-1} c_k z_k^2 + z_n u + \kappa |w_n|^2 z_n^2 + z_n \sum_{k=1}^{n} \Phi_k z_k, \tag{2.129}$$

where

$$\begin{aligned} \Phi_k &= 2\kappa w_k^T w_n + \phi_k, \qquad k = 1, \cdots, n-2 \\ \Phi_{n-1} &= 1 + 2\kappa w_{n-1}^T w_n + \phi_{n-1} \\ \Phi_n &= \phi_n. \end{aligned} \tag{2.130}$$

A control law of the form (2.127) with

$$R_2(x) = \left(c_n + \kappa |w_n|^2 + \sum_{k=1}^{n} \frac{\Phi_k^2}{2c_k} \right)^{-1} > 0, \ c_n > 0, \tag{2.131}$$

results in

$$\dot{V} = -\frac{1}{2} \sum_{k=1}^{n} c_k z_k^2 - \frac{1}{2} \sum_{k=1}^{n} c_k \left(z_k - \frac{\Phi_k}{c_k} z_n \right)^2. \tag{2.132}$$

By Theorem 2.8, the inverse optimal gain assignment problem is solved with the feedback control law

$$u = \alpha_n^*(x) = \beta \alpha_n(x), \qquad \beta \geq 2. \tag{2.133}$$

Remark 2.23 We point out that the choice of α_i, $i = 1, \cdots, n-1$ in (2.123) is the same as in [88], but the control u is chosen differently. While the controller in [88] cancels all the terms inside the bracket in (2.124), the controller presented here does not. As a result, the controller in [88] achieves only attenuation of the effect of the disturbances on z, while the controller here achieves optimality which includes a penalty on u. □

Remark 2.24 The choice of α_i as in (2.123) is not unique. In fact, the iss-clf framework provides more flexibility in choosing the α_i's. We present another choice motivated by the following lemma:

2.6 Design for Strict-Feedback Systems

Lemma 2.25 *For vectors* $y_1, \cdots, y_n$, *the following identity holds:*

$$\kappa \left| \sum_{i=1}^{n} y_i \right|^2 = \sum_{i=1}^{n} \kappa_i |y_i|^2 - \kappa \sum_{\substack{i,j=1 \\ i \neq j}}^{n} \left| \sqrt{\frac{\kappa_i}{\kappa_j}} y_i - \sqrt{\frac{\kappa_j}{\kappa_i}} y_j \right|^2, \quad (2.134)$$

where $\kappa = \left(\sum_{i=1}^{n} \frac{1}{\kappa_i} \right)^{-1}$, $\kappa_i > 0$.

Proof. Since

$$\left| \sum_{i=1}^{n} y_i \right|^2 = \sum_{i=1}^{n} |y_i|^2 + \sum_{\substack{i,j=1 \\ i \neq j}}^{n} 2 y_i^{\mathrm{T}} y_j \quad (2.135)$$

and

$$2 y_i^{\mathrm{T}} y_j = \frac{\kappa_i}{\kappa_j} |y_i|^2 + \frac{\kappa_j}{\kappa_i} |y_j|^2 - \left| \sqrt{\frac{\kappa_i}{\kappa_j}} y_i - \sqrt{\frac{\kappa_j}{\kappa_i}} y_j \right|^2, \quad (2.136)$$

we have that

$$\left| \sum_{i=1}^{n} y_i \right|^2 = \sum_{i=1}^{n} |y_i|^2 + \sum_{\substack{i,j=1 \\ i \neq j}}^{n} \frac{\kappa_i}{\kappa_j} |y_i|^2 + \sum_{\substack{i,j=1 \\ i \neq j}}^{n} \frac{\kappa_j}{\kappa_i} |y_j|^2 - \sum_{\substack{i,j=1 \\ i \neq j}}^{n} \left| \sqrt{\frac{\kappa_i}{\kappa_j}} y_i - \sqrt{\frac{\kappa_j}{\kappa_i}} y_j \right|^2. \quad (2.137)$$

Noting that

$$\sum_{i=1}^{n} |y_i|^2 + \sum_{\substack{i,j=1 \\ i \neq j}}^{n} \frac{\kappa_i}{\kappa_j} |y_i|^2 + \sum_{\substack{i,j=1 \\ i \neq j}}^{n} \frac{\kappa_j}{\kappa_i} |y_j|^2 = \left(\sum_{i=1}^{n} \frac{1}{\kappa_i} \right) \sum_{i=1}^{n} \kappa_i |y_i|^2, \quad (2.138)$$

we get

$$\left| \sum_{i=1}^{n} y_i \right|^2 = \left(\sum_{i=1}^{n} \frac{1}{\kappa_i} \right) \sum_{i=1}^{n} \kappa_i |y_i|^2 - \sum_{\substack{i,j=1 \\ i \neq j}}^{n} \left| \sqrt{\frac{\kappa_i}{\kappa_j}} y_i - \sqrt{\frac{\kappa_j}{\kappa_i}} y_j \right|^2, \quad (2.139)$$

which completes the proof. □

We can now rewrite (2.122) as

$$\begin{aligned}
\dot{V} &= z_n u + \kappa \sum_{i=1}^{n} |w_i z_i|^2 + \sum_{i=1}^{n} \left(x_{i+1} - \sum_{k=1}^{i-1} \frac{\partial \alpha_{i-1}}{\partial x_k} x_{k+1} \right) z_i \\
&\quad - \kappa \sum_{\substack{i,j=1 \\ i \neq j}}^{n} \left| \sqrt{\frac{\kappa_i}{\kappa_j}} w_i z_i - \sqrt{\frac{\kappa_j}{\kappa_i}} w_j z_j \right|^2
\end{aligned}$$

$$= z_n u + \kappa_n |w_n|^2 z_n^2 + z_n z_{n-1} - z_n \sum_{k=1}^{n-1} \frac{\partial \alpha_{n-1}}{\partial x_k} x_{k+1}$$

$$- \kappa \sum_{\substack{i,j=1 \\ i \neq j}}^{n} \left| \sqrt{\frac{\kappa_i}{\kappa_j}} w_i z_i - \sqrt{\frac{\kappa_j}{\kappa_i}} w_j z_j \right|^2$$

$$+ \sum_{i=1}^{n-1} \left[z_{i-1} + \alpha_i + \kappa_i |w_i|^2 z_i - \sum_{k=1}^{i-1} \frac{\partial \alpha_{i-1}}{\partial x_k} x_{k+1} \right] z_i, \qquad (2.140)$$

where $\kappa = \left(\sum_{i=1}^{n} \frac{1}{\kappa_i} \right)^{-1}$, $\kappa_i > 0$. The choice

$$\alpha_i = -z_{i-1} - c_i z_i - \kappa_i |w_i|^2 z_i + \sum_{k=1}^{i-1} \frac{\partial \alpha_{i-1}}{\partial x_k} x_{k+1} \qquad (2.141)$$

results in

$$\dot{V} = -\sum_{k=1}^{n-1} c_k z_k^2 - \kappa \sum_{\substack{i,j=1 \\ i \neq j}}^{n} \left| \sqrt{\frac{\kappa_i}{\kappa_j}} w_i z_i - \sqrt{\frac{\kappa_j}{\kappa_i}} w_j z_j \right|^2 + z_n u + \kappa_n |w_n|^2 z_n^2 + z_n \sum_{k=1}^{n} \Psi_k z_k,$$
$$(2.142)$$

where

$$\Psi_k = \phi_k, \qquad k = 1, \cdots, n-2$$
$$\Psi_{n-1} = 1 + \phi_{n-1} \qquad (2.143)$$
$$\Psi_n = \phi_n.$$

Instead of (2.132), a control law of the form (2.127) with

$$R_2(x) = \left(c_n + \kappa |w_n|^2 + \sum_{k=1}^{n} \frac{\Psi_k^2}{2 c_k} \right)^{-1} > 0 \qquad (2.144)$$

results in

$$\dot{V} = -\frac{1}{2} \sum_{k=1}^{n} c_k z_k^2 - \frac{1}{2} \sum_{k=1}^{n} c_k \left(z_k - \frac{\Psi_k}{c_k} z_n \right)^2 - \kappa \sum_{\substack{i,j=1 \\ i \neq j}}^{n} \left| \sqrt{\frac{\kappa_i}{\kappa_j}} w_i z_i - \sqrt{\frac{\kappa_j}{\kappa_i}} w_j z_j \right|^2.$$
$$(2.145)$$

By Theorem 2.8, the inverse optimal gain assignment problem is solved with the feedback control law $u = \alpha_n^*(x) = \beta \alpha_n(x)$, $\beta \geq 2$. The design in this remark is similar to those in [76] and [63] for steps $i = 1, \ldots, n-1$, but different at step n where the new design selects control of the form (2.127) instead of cancelling the nonlinearities. □

2.7 Performance Estimates

We now give performance bounds on the error state z and control u for the inverse optimal controller designed in Section 2.6.

Theorem 2.26 *In the closed-loop system (2.115), (2.127), the following inequalities hold:*

(i) $$\int_0^\infty \left(2\sum_{k=1}^n c_k z_k^2 + \frac{u^2}{c_n + \kappa |w_n|^2 + \sum_{k=1}^n \frac{\Phi_k^2}{2c_k}} \right) dt \leq \frac{1}{\kappa}\|d\|_2^2 + 2|z(0)|^2 \quad (2.146)$$

(ii) $$|z(t)| \leq \frac{1}{\sqrt{2c\kappa}} \sup_{0 \leq \tau \leq t} |d(\tau)| + |z(0)| e^{-ct/2}, \quad (2.147)$$

where $c = \min_{1 \leq i \leq n} c_i$.

Proof. *(i)* According to Theorem 2.8, the control law $u = \alpha^*(x)$ is optimal with respect to the cost functional ($\beta = \lambda = 2$)

$$J(u) = \sup_d \left\{ \lim_{t \to \infty} \left[4V(x(t)) + \int_0^t \left(2\sum_{k=1}^n c_k z_k^2 + 2\sum_{k=1}^n c_k \left(z_k - \frac{\Phi_k}{c_k} z_n \right)^2 \right. \right. \right.$$
$$\left. \left. \left. + \frac{u^2}{c_n + \kappa|w_n|^2 + \sum_{k=1}^n \frac{\Phi_k^2}{2c_k}} - \frac{1}{\kappa}|d|^2 \right) d\tau \right] \right\} \quad (2.148)$$

with a value function

$$J^* = 2|z|^2. \quad (2.149)$$

Therefore,

$$\int_0^\infty \left[2\sum_{k=1}^n c_k z_k^2 + \frac{u^2}{c_n + \kappa|w_n|^2 + \sum_{k=1}^n \frac{\Phi_k^2}{2c_k}} - \frac{1}{\kappa}|d|^2 \right] dt$$

$$\leq \int_0^\infty \left[2\sum_{k=1}^n c_k z_k^2 + 2\sum_{k=1}^n c_k \left(z_k - \frac{\Phi_k}{c_k} z_n \right)^2 + \frac{u^2}{c_n + \kappa|w_n|^2 + \sum_{k=1}^n \frac{\Phi_k^2}{2c_k}} - \frac{1}{\kappa}|d|^2 \right] dt$$

$$\leq J^* = 2|z(0)|^2, \quad (2.150)$$

which yields (2.146).

(ii) Differentiating $\frac{1}{2}|z|^2$ along the solutions of (2.115), noting (2.132), we have

$$\frac{d}{dt}\left(\frac{1}{2}|z|^2\right) = -\frac{1}{2}\sum_{k=1}^n c_k z_k^2 - \frac{1}{2}\sum_{k=1}^n c_k \left(z_k - \frac{\Phi_k}{c_k} z_n \right)^2$$

$$-\kappa \left|\sum_{k=1}^{n} w_k z_k\right|^2 + \left(\sum_{k=1}^{n} w_k z_k\right)^T d$$
$$\leq -\frac{c}{2}|z|^2 + \frac{1}{4\kappa}|d|^2. \tag{2.151}$$

By the comparison principle, we get (2.147). □

Remark 2.27 If the control law is chosen as in Remark 2.24, the cost functional is

$$J(u) = \sup_{d} \left\{ \lim_{t \to \infty} \left[4V(x(t)) + 2\int_0^t \left(\sum_{k=1}^{n} c_k z_k^2 + \sum_{k=1}^{n} c_k \left(z_k - \frac{\Psi_k}{c_k} z_n\right)^2 \right. \right. \right.$$
$$+ 2\kappa \sum_{\substack{i,j=1 \\ i \neq j}}^{n} \left| \sqrt{\frac{\kappa_i}{\kappa_j}} w_i z_i - \sqrt{\frac{\kappa_j}{\kappa_i}} w_j z_j \right|^2$$
$$\left. \left. \left. + \frac{u^2}{2\left(c_n + \kappa |w_n|^2 + \sum_{k=1}^{n} \frac{\Psi_k^2}{2 c_k}\right)} - \frac{1}{2\kappa}|d|^2 \right) d\tau \right] \right\}. \tag{2.152}$$

Instead of (2.146), we have the performance bound

$$\int_0^\infty \left(2 \sum_{k=1}^{n} c_k z_k^2 + \frac{u^2}{c_n + \kappa |w_n|^2 + \sum_{k=1}^{n} \frac{\Psi_k^2}{2 c_k}} \right) dt \leq \frac{1}{\kappa}\|d\|_2^2 + 2|z(0)|^2. \tag{2.153}$$

The control law from Remark 2.24 also achieves an ISS bound as in (2.147). □

Notes and References

Disturbance attenuation is a subset of a more general *robust nonlinear control problem*. An in-depth coverage of robust nonlinear control is given in the book of Freeman and Kokotović [32]. The nonlinear disturbance attenuation problem has been widely studied as a "nonlinear $\mathcal{H}_\infty$" or a differential game problem in the books of van der Schaft [112], Başar and Bernhard [7] and Isidori [44], and in a number of papers [6, 17, 41, 43, 45, 46, 48, 49, 61, 73, 102, 109, 110, 111]. The perennial obstacle to implementation of these controllers is the need to solve Hamilton-Jacobi-Isaacs partial differential equations. For this reason, parallel efforts on disturbance attenuation have proceeded in the framework of Sontag's input-to-state stability (ISS) [100] where a number of constructive designs have been reported [32, 54, 63, 88, 93, 101, 105]. Theorem 2.6 was established in [63] in the context of modular adaptive control. Krstić and Li [64] showed that input-to-state stabilizability is equivalent to the

Notes and References

solvability of a differential game problem similar to, but more general than, the nonlinear $\mathcal{H}_\infty$ problem. This chapter is based on that reference.

Even though there were notable prior efforts [83, 34], it is fair to say that inverse optimality was introduced into nonlinear control via Freeman's robust control Lyapunov functions [32]. The interest in inverse, rather that direct optimality, is because (1) not every direct problem has a solution (as illustrated by the example at the beginning of Section 2.3) and (2) inverse optimality guarantees that the controller remains input-to-state stabilizing in the presence of a certain class of input unmodeled dynamics which do not have to be small in the $\mathcal{H}_\infty$ sense, do not have to be linear, and do not even have to be ISS. Efforts on input unmodeled dynamics have been intensive over the last few years, starting with Krstić, Sun and Kokotović [65] and followed by Jiang and Pomet [53], Praly and Wang [93], and Jiang and Mareels [52]. Sepulchre, Janković and Kokotović [96] were the first in the nonlinear setting to quantify stability margins to input unmodeled dynamics. They did this by using inverse optimality and passivity concepts. The result presented in Section 2.4, originally reported in Krstić and Li [64], is an extension of their result to the case with disturbances. Unfortunately, like in the linear case (even) without disturbances [1], the class of allowable input dynamics does not include those that increase relative degree (and thus reduce the control authority at higher frequencies) such as, e.g., $1/(1+\mu s)$, which are typical actuator dynamics.

The constructive character of the inverse optimal approach is illustrated in Sections 2.5–2.7, and is based on [64]. Prior disturbance attenuation constructions for strict-feedback systems have been reported by Marino, Respondek, van der Schaft, and Tomei [76], Isidori [44], Krstić, Kanellakopoulos, and Kokotović [63], and Pan and Başar [88], but without a cost on the control effort. A new paper by Ezal, Pan, and Kokotovic [24] achieves not only inverse optimality but also local *direct* optimality with a single controller.

Part II
STOCHASTIC SYSTEMS

Chapter 3

Stochastic Stability and Regulation

In this chapter we develop a stochastic analog of the results in Chapter 1. In the systems that we study here the noise enters in a manner that does not destroy the equilibrium at the origin. This excludes linear systems where the noise usually enters in an additive fashion. Linear systems are covered by the results in the next chapter.

This chapter lays the foundations for stochastic stability analysis (upon which the results of both this and Chapters 4 and 6 are built), solves the global stabilization (in probability) problem, and achieves inverse optimality. This is the only chapter that presents output-feedback results—as an example of what is achievable with the methods in this book.

3.1 Stochastic Lyapunov and LaSalle Like Theorems

Consider the nonlinear stochastic system

$$dx = f(x)dt + g(x)dw, \qquad (3.1)$$

where $x \in \mathbb{R}^n$ is the state, w is an r-dimensional independent standard Wiener process, and $f : \mathbb{R}^n \to \mathbb{R}^n$ and $g : \mathbb{R}^n \to \mathbb{R}^{n \times r}$ are locally Lipschitz and $f(0) = 0$, $g(0) = 0$.

Definition 3.1 *The equilibrium $x = 0$ of the system (3.1) is*

- *globally stable in probability if $\forall \epsilon > 0$ there exists a class $\mathcal{K}$ function $\gamma(\cdot)$ such that*

$$P\{|x(t)| < \gamma(|x_0|)\} \geq 1 - \epsilon, \qquad \forall t \geq 0, \forall x_0 \in \mathbb{R}^n \backslash \{0\}, \qquad (3.2)$$

- globally asymptotically stable in probability *if* $\forall \epsilon > 0$ *there exists a class* $\mathcal{KL}$ *function* $\beta(\cdot, \cdot)$ *such that*

$$P\{|x(t)| < \beta(|x_0|, t)\} \geq 1 - \epsilon, \qquad \forall t \geq 0, \forall x_0 \in \mathbb{R}^n \setminus \{0\}. \tag{3.3}$$

Theorem 3.2 *Consider system (3.1) and suppose there exists a C^2 function $V(x)$ and class $\mathcal{K}_\infty$ functions α_1 and α_2, such that*

$$\alpha_1(|x|) \leq V(x) \leq \alpha_2(|x|) \tag{3.4}$$

$$\mathcal{L}V(x) = \frac{\partial V}{\partial x} f(x) + \frac{1}{2} \text{Tr}\left\{ g^T \frac{\partial^2 V}{\partial x^2} g \right\} \leq -W(x), \tag{3.5}$$

where $W(x)$ is continuous and nonnegative. Then the equilibrium $x = 0$ is globally stable in probability and

$$P\left\{ \lim_{t \to \infty} W(x) = 0 \right\} = 1. \tag{3.6}$$

Proof. Since $\mathcal{L}V \leq 0$ and V is radially unbounded, there exists globally a unique solution [59, pp. 84, Theorem 4.1] with probability one (that is, the probability of finite escape is zero and the probability that two solutions starting from the same initial condition are different is zero). Again, since $\mathcal{L}V \leq 0$, $V(x(t))$ is a supermartingale, so $EV(t) \leq V_0$. Applying Chebyshev's inequality, for any class $\mathcal{K}_\infty$ function $\delta(\cdot)$, we have

$$P\{V \geq \delta(V_0)\} \leq \frac{EV}{\inf_{V \geq \delta(V_0)} V} \leq \frac{V_0}{\delta(V_0)}, \qquad \forall V_0 \neq 0 \tag{3.7}$$

so,

$$P\{V < \delta(V_0)\} \geq 1 - \frac{V_0}{\delta(V_0)}, \qquad \forall V_0 \neq 0 \tag{3.8}$$

Denote $\rho = \alpha_1^{-1} \circ \delta \circ \alpha_2$, then $V(t) < \delta(V_0)$ implies $|x(t)| < \rho(|x_0|)$, and thus

$$P\{|x(t)| < \rho(|x_0|)\} \geq 1 - \frac{V_0}{\delta(V_0)}, \qquad \forall V_0 \neq 0 \tag{3.9}$$

For a given $\epsilon > 0$, choose $\delta(\cdot)$ such that

$$\delta(V_0) \geq \frac{V_0}{\epsilon}. \tag{3.10}$$

Then we have

$$P\{|x(t)| < \rho(|x_0|)\} \geq 1 - \epsilon, \qquad \forall t \geq 0, \forall x_0 \in \mathbb{R}^n \setminus \{0\} \tag{3.11}$$

and the global stability in probability is proved.

3.1 STOCHASTIC LYAPUNOV AND LASALLE LIKE THEOREMS

Now we begin to prove the convergence of $W(x)$. By Chebyshev's inequality,

$$P\left\{\sup_{t\geq t_0} |x(t)| \geq r\right\} \leq P\left\{\sup_{t\geq t_0} V(t) \geq \alpha_1(r)\right\}$$
$$\leq \frac{\sup_{t\geq t_0} EV(t)}{\alpha_1(r)}$$
$$\leq \frac{V_0}{\alpha_1(r)}, \quad \forall r > 0 \quad (3.12)$$

that is,

$$p = P\left\{\sup_{t\geq t_0} |x(t)| < r\right\} \geq 1 - \frac{V_0}{\alpha_1(r)}, \quad \forall r > 0. \quad (3.13)$$

For a trajectory starting from x_0 and bounded by $r > 0$, there exists a time $\tau(x_0)$ such that one of the following three mutually exclusive possibilities is satisfied:

1. for any $\epsilon > 0$, there exists $T > \tau(x_0)$ such that $W(x(t)) < \epsilon, \forall t \geq T$.

2. $\exists \epsilon > 0$ such that $W(x(t)) > \epsilon, \forall t \geq \tau(x_0)$.

3. $\exists \epsilon_0 > 0$ such that, $\forall \epsilon' \in (0, \epsilon_0)$, $W(x(t))$ jumps from below ϵ' to above $2\epsilon'$ and back infinitely many times after any $t \geq \tau(x_0)$.

Denote the three events by A_1, A_2, A_3, and

$$p_i = P\left\{A_i \Big| \sup_{t\geq t_0} |x(t)| < r\right\}, \quad i = 1, 2, 3. \quad (3.14)$$

Then $p_1 + p_2 + p_3 = 1$. From (3.1) we compute,

$$E\left\{\sup_{t\leq s\leq t+h} |x(s) - x(t)|^2 \Big| \sup_{t\geq t_0} |x(t)| < r\right\}$$
$$= E\left\{\sup_{t\leq s\leq t+h} \left|\int_t^s f(x)d\tau + \int_t^s g(x)dw\right|^2 \Big| \sup_{t\geq t_0} |x(t)| < r\right\}$$
$$\leq 2E\left\{\sup_{t\leq s\leq t+h} \left|\int_t^s f(x)d\tau\right|^2 \Big| \sup_{t\geq t_0} |x(t)| < r\right\}$$
$$+ 2E\left\{\sup_{t\leq s\leq t+h} \left|\int_t^s g(x)dw\right|^2 \Big| \sup_{t\geq t_0} |x(t)| < r\right\}$$
$$\leq \rho_1(r)h^2 + 2E\left\{\sup_{t\leq s\leq t+h} \left|\int_t^s g(x)dw\right|^2 \Big| \sup_{t\geq t_0} |x(t)| < r\right\}, \quad (3.15)$$

where the last inequality follows from the continuity of $f(x)$ and $\rho_1(r)$ is a class $\mathcal{K}$ function. Applying the continuous parameter version of Theorem 3.4

in Doob [18, pp. 317], and then Lemma 3.5 in Øksendal [86] (the Itô isometry) to the second term, we have

$$
2E\left\{\sup_{t\leq s\leq t+h}\left|\int_t^s g(x)dw\right|^2 \Big| \sup_{t\geq t_0}|x(t)| < r\right\}
$$
$$
\leq 8E\left\{\left|\int_t^{t+h} g(x)dw\right|^2 \Big| \sup_{t\geq t_0}|x(t)| < r\right\}
$$
$$
= 8\int_t^{t+h} E\left\{|g(x)|^2 \Big| \sup_{t\geq t_0}|x(t)| < r\right\} d\tau
$$
$$
\leq \rho_2(r)h, \tag{3.16}
$$

where the last inequality follows from the continuity of $g(x)$ and $\rho_2(r)$ is a class $\mathcal{K}$ function. So, we have

$$
E\left\{\sup_{t\leq s\leq t+h}|x(s)-x(t)|^2 \Big| \sup_{t\geq t_0}|x(t)| < r\right\}
$$
$$
\leq \rho_1(r)h^2 + \rho_2(r)h
$$
$$
\leq \rho(r)\eta(h) \tag{3.17}
$$

for all $r > 0$ and $h > 0$, where ρ and η are class $\mathcal{K}$ functions. Since $W(x)$ is continuous, from Corollary A.5 of Randy Freeman [32], there exists a class $\mathcal{K}$ function γ such that

$$
E\left\{\sup_{t\leq s\leq t+h}|W(x(s))-W(x(t))|^2 \Big| \sup_{t\geq t_0}|x(t)| < r\right\}
$$
$$
\leq \left\{\sup_{t\leq s\leq t+h}\gamma\left(|x(s)-x(t)|^2\right) \Big| \sup_{t\geq t_0}|x(t)| < r\right\}
$$
$$
\leq \gamma(\rho(r)\eta(h)). \tag{3.18}
$$

By Chebyshev's inequality,

$$
P\left\{\sup_{t\leq s\leq t+h}|W(x(s))-W(x(t))| > \epsilon' \Big| \sup_{t\geq t_0}|x(t)| < r\right\} \leq \frac{\gamma(\rho(r)\eta(h))}{\epsilon'^2}. \tag{3.19}
$$

For every $r > 0$, we can find an $h^*(r,\epsilon') > 0$, such that, for all $h \in (0, h^*]$,

$$
P\left\{\sup_{t\leq s\leq t+h}|W(x(s))-W(x(t))| \leq \epsilon' \Big| \sup_{t\geq t_0}|x(t)| < r\right\}
$$
$$
\geq \left(1 - \frac{\gamma(\rho(r)\eta(h))}{\epsilon'^2}\right) > 0. \tag{3.20}
$$

Then

$$
EV(t) = V(t_0) + \int_{t_0}^t E\mathcal{L}V\,ds
$$

3.1 STOCHASTIC LYAPUNOV AND LASALLE LIKE THEOREMS

$$\begin{aligned}
&= V(t_0) - \int_{t_0}^{t} EW(x)ds \\
&\leq V(t_0) - \int_{t_0}^{t} \inf_{\sup_{t \geq t_0}|x|>r}\{W(x)\}(1-p)ds - \int_{t_0}^{t} \inf_{W(x)<\epsilon}\{W(x)\}p_1 p\, ds \\
&\quad - \int_{t_0}^{t} \inf_{W(x)\geq \epsilon}\{W(x)\}p_2 p\, ds - p_3 phN(t)\epsilon'\left(1 - \frac{\gamma(\rho(r)\eta(h))}{\epsilon'^2}\right) \\
&\leq V(t_0) - \epsilon p_2 p(t-t_0) - p_3 phN(t)\epsilon'\left(1 - \frac{\gamma(\rho(r)\eta(h))}{\epsilon'^2}\right), \quad (3.21)
\end{aligned}$$

where $N(t)$ is the number of jumps from $\tau(x_0)$ to t, and $N(t) \to \infty$ as $t \to \infty$. If either p_2 or p_3 is positive, $EV(t)$ will become negative when t is large enough, which is in contradiction with the positiveness of $V(t)$. Thus, $p_2 = p_3 = 0$ and $p_1 = 1$. This implies

$$\begin{aligned}
&P\left\{\forall \epsilon > 0, \exists T \text{ such that } W(x(t)) < \epsilon \; \forall t > T, \text{ and } \sup_{t \geq t_0}|x| < r\right\} \\
&= P\left\{\forall \epsilon > 0, \exists T \text{ such that } W(x(t)) < \epsilon \; \forall t > T \,\Big|\, \sup_{t \geq t_0}|x| < r\right\} P\left\{\sup_{t \geq t_0}|x| < r\right\} \\
&= p_1 p \\
&\geq 1 - \frac{V(x_0)}{\alpha_1(r)}. \quad (3.22)
\end{aligned}$$

Letting $r \to \infty$, we have

$$P\left\{\lim_{t \to \infty} W(x(t)) = 0\right\} = 1. \quad (3.23)$$

$\square$

Theorem 3.3 *Consider system (3.1) and suppose there exists a C^2 function $V(x)$, class $\mathcal{K}_\infty$ functions α_1 and α_2, and a class $\mathcal{K}$ function α_3, such that*

$$\alpha_1(|x|) \leq V(x) \leq \alpha_2(|x|), \quad (3.24)$$

$$\mathcal{L}V(x) = \frac{\partial V}{\partial x}f(x) + \frac{1}{2}\mathrm{Tr}\left\{g^{\mathrm{T}}\frac{\partial^2 V}{\partial x^2}g\right\} \leq -\alpha_3(|x|). \quad (3.25)$$

Then the equilibrium $x = 0$ is globally asymptotically stable in probability.

Proof. This theorem is a direct corollary of Theorem 3.2 due to the fact that $W(x) = \alpha_3(|x|)$ is positive definite and implies $P\{\lim_{t \to \infty}|x(t)| = 0\} = 1$. Combining this with

$$P\{|x(t)| < \rho(|x_0|)\} \geq 1 - \epsilon, \quad \forall t \geq 0, \forall x_0 \in \mathbb{R}^n \setminus \{0\} \quad (3.26)$$

we have

$$P\left\{|x(t)| < \rho(|x_0|) \text{ and } \lim_{t\to\infty} |x(t)| = 0\right\}$$
$$= P\left\{|x(t)| < \rho(|x_0|)\right\} P\left\{\lim_{t\to\infty} |x(t)| = 0\right\}$$
$$\geq 1 - \epsilon, \quad \forall t \geq 0, \ \forall x_0 \in \mathbb{R}^n \setminus \{0\}. \tag{3.27}$$

The expression in the first line implies the existence of a class $\mathcal{KL}$ function β such that (3.3) holds. □

3.2 Stabilization in Probability via Backstepping

In this section we deal with *strict-feedback* systems driven by white noise. This class of systems is given by the following nonlinear stochastic differential equations:

$$dx_i = x_{i+1} dt + \varphi_i(\bar{x}_i)^T dw, \quad i = 1, \cdots, n-1 \tag{3.28}$$
$$dx_n = u\, dt + \varphi_n(\bar{x}_n)^T dw, \tag{3.29}$$

where $\bar{x}_i = [x_1, \cdots, x_i]^T$, $\varphi_i(\bar{x}_i)$ are r-vector-valued smooth functions with

$$\varphi_i(0) = 0, \tag{3.30}$$

and w is an independent r-dimensional *standard* Wiener process.

In the standard backstepping approach for *deterministic* systems described in Sections 1.2 and 2.6 (where dw/dt would be a bounded deterministic disturbance), a sequence of stabilizing functions $\alpha_i(\bar{x}_i)$ would be constructed recursively to build a Lyapunov function of the form

$$V = \sum_{i=1}^n \frac{1}{2} z_i^2, \tag{3.31}$$

where the error variables z_i are given by

$$z_i = x_i - \alpha_{i-1}(\bar{x}_{i-1}). \tag{3.32}$$

The Lyapunov design for stochastic systems is *much* more difficult because of the term $\frac{1}{2}\text{Tr}\left\{g^T \frac{\partial^2 V}{\partial x^2} g\right\}$ in (3.5), and it cannot be carried out using (3.31). Our approach in dealing with the inadequacy of quadratic Lyapunov functions is to employ *quartic* Lyapunov functions

$$V = \sum_{i=1}^n \frac{1}{4} z_i^4. \tag{3.33}$$

3.2 STABILIZATION IN PROBABILITY VIA BACKSTEPPING

We start by several important preparatory comments. Since $\varphi_i(0) = 0$, the α_i's will vanish at $\bar{x}_i = 0$, as well as at $\bar{z}_i = 0$, where $\bar{z}_i = [z_1, \cdots, z_i]^{\mathrm{T}}$. Thus, by the mean value theorem, $\alpha_i(\bar{x}_i)$ can be expressed as

$$\alpha_i(\bar{x}_i) = \sum_{l=1}^{i} z_l \alpha_{il}(\bar{x}_i), \qquad (3.34)$$

where $\alpha_{il}(\bar{x}_i)$ are smooth functions. We can now write $\varphi_i(\bar{x}_i)$ as

$$\begin{aligned} \varphi_i(\bar{x}_i) &= \sum_{k=1}^{i} x_k \varphi_{ik}(\bar{x}_i) \\ &= \sum_{k=1}^{i} (z_k + \alpha_{k-1}) \varphi_{ik}(\bar{x}_i) \\ &= \sum_{k=1}^{i} z_k \psi_{ik}(\bar{x}_i), \end{aligned} \qquad (3.35)$$

where $\varphi_{ik}(\bar{x}_i)$ and $\psi_{ik}(\bar{x}_i)$ are smooth functions. Then the system (3.28), (3.29) can be written as

$$dx_i = x_{i+1}dt + \left(\sum_{k=1}^{i} z_k \psi_{ik}(\bar{x}_i)^{\mathrm{T}}\right) dw, \qquad i = 1, \cdots, n-1 \qquad (3.36)$$

$$dx_n = u\,dt + \left(\sum_{k=1}^{n} z_k \psi_{nk}(\bar{x}_n)^{\mathrm{T}}\right) dw. \qquad (3.37)$$

Now, we are ready to start the backstepping design procedure. According to Itô's differentiation rule [86], we have

$$\begin{aligned} dz_i &= d\left(x_i - \alpha_{i-1}\right) \\ &= \left(x_{i+1} - \sum_{l=1}^{i-1} \frac{\partial \alpha_{i-1}}{\partial x_l} x_{l+1} - \frac{1}{2} \sum_{p,q=1}^{i-1} \frac{\partial^2 \alpha_{i-1}}{\partial x_p \partial x_q} \varphi_p^{\mathrm{T}} \varphi_q\right) dt \\ &\quad + \left(\varphi_i^{\mathrm{T}} - \sum_{l=1}^{i-1} \frac{\partial \alpha_{i-1}}{\partial x_l} \varphi_l^{\mathrm{T}}\right) dw, \qquad i = 1, \cdots, n, \end{aligned} \qquad (3.38)$$

where $x_{n+1} = u$. As we announced previously, we employ a Lyapunov function of the form

$$V(z) = \sum_{i=1}^{n} \frac{1}{4} z_i^4, \qquad (3.39)$$

and set out to select the functions $\alpha_i(\bar{x}_i)$ to make $\mathcal{L}V(x)$ negative definite. Along the solutions of (3.38), we have

$$\mathcal{L}V = z_n^3 \left(u - \sum_{l=1}^{n-1} \frac{\partial \alpha_{n-1}}{\partial x_l} x_{l+1} - \frac{1}{2} \sum_{p,q=1}^{n-1} \frac{\partial^2 \alpha_{n-1}}{\partial x_p \partial x_q} \varphi_p^{\mathrm{T}} \varphi_q\right)$$

$$+ \sum_{i=1}^{n-1} z_i^3 \left(x_{i+1} - \sum_{l=1}^{i-1} \frac{\partial \alpha_{i-1}}{\partial x_l} x_{l+1} - \frac{1}{2} \sum_{m,n=1}^{i-1} \frac{\partial^2 \alpha_{i-1}}{\partial x_m \partial x_n} \varphi_m^T \varphi_n \right)$$

$$+ \frac{3}{2} \sum_{i=1}^{n} z_i^2 \left(\varphi_i - \sum_{l=1}^{i-1} \frac{\partial \alpha_{i-1}}{\partial x_l} \varphi_l \right)^T \left(\varphi_i - \sum_{l=1}^{i-1} \frac{\partial \alpha_{i-1}}{\partial x_l} \varphi_l \right)$$

$$= z_n^3 \left(u - \sum_{l=1}^{n-1} \frac{\partial \alpha_{n-1}}{\partial x_l} x_{l+1} - \frac{1}{2} \sum_{p,q=1}^{n-1} \frac{\partial^2 \alpha_{n-1}}{\partial x_p \partial x_q} \varphi_p^T \varphi_q \right) + \sum_{i=1}^{n-1} z_i^3 z_{i+1}$$

$$+ \sum_{i=1}^{n-1} z_i^3 \left(\alpha_i - \sum_{l=1}^{i-1} \frac{\partial \alpha_{i-1}}{\partial x_l} x_{l+1} - \frac{1}{2} \sum_{p,q=1}^{i-1} \frac{\partial^2 \alpha_{i-1}}{\partial x_p \partial x_q} \varphi_p^T \varphi_q \right)$$

$$+ \frac{3}{2} \sum_{i=1}^{n} z_i^2 \left(\varphi_i - \sum_{l=1}^{i-1} \frac{\partial \alpha_{i-1}}{\partial x_l} \varphi_l \right)^T \left(\varphi_i - \sum_{l=1}^{i-1} \frac{\partial \alpha_{i-1}}{\partial x_l} \varphi_l \right), \quad (3.40)$$

where the second equality comes from substituting $x_i = z_i + \alpha_{i-1}$. The first and the third term can be handled by choosing u and α_i which cancel the summations. To handle the second and the fourth term, we need Young's inequality (Corollary A.3)

$$xy \leq \frac{\epsilon^p}{p} |x|^p + \frac{1}{q \epsilon^q} |y|^q, \quad (3.41)$$

where $\epsilon > 0$, the constants $p > 1$ and $q > 1$ satisfy $(p-1)(q-1) = 1$, and $(x,y) \in \mathbb{R}^2$. Applying (3.41) to the second term in (3.40), using $p = \frac{4}{3}$, $q = 4$ and $\epsilon_i > 0$ for each i, we have

$$\sum_{i=1}^{n-1} z_i^3 z_{i+1} \leq \frac{3}{4} \sum_{i=1}^{n-1} \epsilon_i^{\frac{4}{3}} z_i^4 + \sum_{i=2}^{n} \frac{1}{4\epsilon_{i-1}^4} z_i^4. \quad (3.42)$$

Substituting (3.42) into (3.40), and incorporating into the first and the third term, the terms originated in (3.42) are easily cancelled by u and α_i.

However, the last term in (3.40) remains difficult to handle. We try to arrange it in the form

$$\sum_{i=1}^{n} z_i^\gamma \eta_i(\bar{x}_i), \quad \gamma \geq 3, \quad (3.43)$$

so that it also can be combined into the first and the third term in (3.40), and be cancelled by α_i and u. Substituting (3.35) into the last term in (3.40) yields

$$\frac{3}{2} \sum_{i=1}^{n} z_i^2 \left(\varphi_i - \sum_{l=1}^{i-1} \frac{\partial \alpha_{i-1}}{\partial x_l} \varphi_l \right)^T \left(\varphi_i - \sum_{l=1}^{i-1} \frac{\partial \alpha_{i-1}}{\partial x_l} \varphi_l \right)$$

$$= \frac{3}{2} \sum_{i=1}^{n} z_i^2 \left(\sum_{k=1}^{i} z_k \psi_{ik} - \sum_{l=1}^{i-1} \frac{\partial \alpha_{i-1}}{\partial x_l} \sum_{k=1}^{l} z_k \psi_{lk} \right)^T \left(\sum_{k=1}^{i} z_k \psi_{ik} - \sum_{l=1}^{i-1} \frac{\partial \alpha_{i-1}}{\partial x_l} \sum_{k=1}^{l} z_k \psi_{lk} \right)$$

3.2 STABILIZATION IN PROBABILITY VIA BACKSTEPPING

$$= \frac{3}{2}\sum_{i=1}^{n} z_i^2 \left(\sum_{k=1}^{i} z_k\psi_{ik} - \sum_{k=1}^{i-1}\sum_{l=1}^{i-1} z_k \frac{\partial \alpha_{i-1}}{\partial x_l}\psi_{lk}\right)^{\mathrm{T}} \left(\sum_{k=1}^{i} z_k\psi_{ik} - \sum_{k=1}^{i-1}\sum_{l=1}^{i-1} z_k \frac{\partial \alpha_{i-1}}{\partial x_l}\psi_{lk}\right)$$

$$= \frac{3}{2}\sum_{i=1}^{n} \left\{ z_i^2 \left[\sum_{k=1}^{i} z_k \left(\psi_{ik} - \sum_{l=k}^{i-1} \frac{\partial \alpha_{i-1}}{\partial x_l}\psi_{lk}\right)\right]^{\mathrm{T}} \left[\sum_{k=1}^{i} z_k \left(\psi_{ik} - \sum_{l=k}^{i-1} \frac{\partial \alpha_{i-1}}{\partial x_l}\psi_{lk}\right)\right] \right\}$$

$$= \frac{3}{2}\sum_{i=1}^{n} z_i^2 \left(\sum_{k=1}^{i} z_k\beta_{ik}\right)^{\mathrm{T}} \left(\sum_{k=1}^{i} z_k\beta_{ik}\right), \qquad (3.44)$$

where $\partial\alpha_{i-1}/\partial x_i = 0$ and

$$\beta_{ik}(\bar{x}_i) = \psi_{ik} - \sum_{l=k}^{i-1} \frac{\partial \alpha_{i-1}}{\partial x_l}\psi_{lk}, \qquad k = 1, \cdots, i. \qquad (3.45)$$

The second equality in (3.44) comes from changing the order of summations. It is obvious that β_{ik} depend only on $\bar{x}_i$. The next major step is to separate different z_i from each other, so every term can be handled by the proper α_i. Hence, we rearrange (3.44) as

$$\frac{3}{2}\sum_{i=1}^{n} z_i^2 \left(\sum_{k=1}^{i} z_k\beta_{ik}\right)^{\mathrm{T}} \left(\sum_{k=1}^{i} z_k\beta_{ik}\right)$$

$$= \frac{3}{2}\sum_{i=1}^{n}\sum_{j=1}^{r} \left[z_i^4 \beta_{iij}^2 + 2z_i^3 \beta_{iij} \sum_{k=1}^{i-1} z_k\beta_{ikj} + z_i^2 \left(\sum_{k=1}^{i-1} z_k\beta_{ikj}\right)^2\right]$$

$$= \frac{3}{2}z_n^4 \beta_{nn}^{\mathrm{T}}\beta_{nn} + 3z_n^3 \beta_{nn}^{\mathrm{T}} \sum_{k=1}^{n-1} z_k\beta_{nk} + \frac{3}{2}\sum_{i=1}^{n-1} \left(z_i^4 \beta_{ii}^{\mathrm{T}}\beta_{ii} + 2z_i^3 \beta_{ii}^{\mathrm{T}} \sum_{k=1}^{i-1} z_k\beta_{ik}\right)$$

$$+ \frac{3}{2}\sum_{i=1}^{n}\sum_{j=1}^{r} z_i^2 \left(\sum_{k=1}^{i-1} z_k\beta_{ikj}\right)^2, \qquad (3.46)$$

where β_{ikj} is the jth component of the vector β_{ik}. The first three terms in (3.46) are already in the desired form. Now we concentrate on the last term in (3.46):

$$\frac{3}{2}\sum_{i=1}^{n}\sum_{j=1}^{r} z_i^2 \left(\sum_{k=1}^{i-1} z_k\beta_{ikj}\right)^2$$

$$= \frac{3}{2}\sum_{i=1}^{n}\sum_{j=1}^{r} z_i^2 \left(\sum_{k=1}^{i-1} z_k\beta_{ikj}\right)\left(\sum_{l=1}^{i-1} z_l\beta_{ilj}\right)$$

$$= \frac{3}{2}\sum_{i=1}^{n}\sum_{j=1}^{r}\sum_{k=1}^{i-1}\sum_{l=1}^{i-1} z_i^2 \beta_{ikj}\beta_{ilj} z_k z_l$$

$$\leq \frac{3}{4}\sum_{i=1}^{n}\sum_{j=1}^{r}\sum_{k=1}^{i-1}\sum_{l=1}^{i-1} \left(\frac{1}{\epsilon_{ikl}^2} z_i^4 \beta_{ikj}^2 \beta_{ilj}^2 + \epsilon_{ikl}^2 z_k^2 z_l^2\right)$$

$$\leq \frac{3}{4}\sum_{i=1}^{n}\sum_{j=1}^{r}\sum_{k=1}^{i-1}\sum_{l=1}^{i-1}\frac{1}{\epsilon_{ikl}^2}z_i^4\beta_{ikj}^2\beta_{ilj}^2 + \frac{3}{4}\sum_{i=1}^{n}\sum_{j=1}^{r}\sum_{k=1}^{i-1}z_k^4\left(\sum_{l=1}^{i-1}\epsilon_{ikl}^2\right)$$

$$= \frac{3}{4}z_n^4\left(\sum_{j=1}^{r}\sum_{k=1}^{n-1}\sum_{l=1}^{n-1}\frac{1}{\epsilon_{nkl}^2}\beta_{nkj}^2\beta_{nlj}^2\right) + \frac{3}{4}\sum_{i=1}^{n-1}z_i^4\left(\sum_{j=1}^{r}\sum_{k=1}^{i-1}\sum_{l=1}^{i-1}\frac{1}{\epsilon_{ikl}^2}\beta_{ikj}^2\beta_{ilj}^2\right)$$

$$+\frac{3r}{4}\sum_{i=1}^{n-1}z_i^4\left(\sum_{k=i+1}^{n}\sum_{l=1}^{k-1}\epsilon_{kil}^2\right). \qquad (3.47)$$

The inequalities above are obtained by applying Young's inequalities:

$$z_i^2\beta_{ikj}\beta_{ilj}z_kz_l \leq \frac{1}{2\epsilon_{ikl}^2}z_i^4\beta_{ikj}^2\beta_{ilj}^2 + \frac{\epsilon_{ikl}^2}{2}z_k^2z_l^2 \qquad (3.48)$$

$$z_k^2z_l^2 \leq \frac{1}{2}z_k^4 + \frac{1}{2}z_l^4, \qquad (3.49)$$

where $\epsilon_{ikl} = \epsilon_{ilk}$. The final expression in (3.47) is obtained by separating the z_n from the other z_i's in the first term and changing the order of summations in the second term.

Now, substituting (3.42), (3.46) and (3.47) into (3.40), we get

$$\mathcal{L}V \leq z_n^3\left(u - \sum_{l=1}^{n-1}\frac{\partial\alpha_{n-1}}{\partial x_l}x_{l+1} - \frac{1}{2}\sum_{p,q=1}^{n-1}\frac{\partial^2\alpha_{n-1}}{\partial x_p\partial x_q}\varphi_p^T\varphi_q\right) + \frac{3}{4}\sum_{i=1}^{n-1}\epsilon_i^{\frac{4}{3}}z_i^4$$

$$+\sum_{i=2}^{n}\frac{1}{4\epsilon_{i-1}^4}z_i^4 + \sum_{i=1}^{n-1}z_i^3\left(\alpha_i - \sum_{l=1}^{i-1}\frac{\partial\alpha_{i-1}}{\partial x_l}x_{l+1} - \frac{1}{2}\sum_{p,q=1}^{i-1}\frac{\partial^2\alpha_{i-1}}{\partial x_p\partial x_q}\varphi_p^T\varphi_q\right)$$

$$+\frac{3}{2}z_n^4\beta_{nn}^T\beta_{nn} + 3z_n^3\beta_{nn}^T\sum_{k=1}^{n-1}z_k\beta_{nk} + \frac{3}{2}\sum_{i=1}^{n-1}\left(z_i^4\beta_{ii}^T\beta_{ii} + 2z_i^3\beta_{ii}^T\sum_{k=1}^{i-1}z_k\beta_{ik}\right)$$

$$+\frac{3}{4}z_n^4\sum_{j=1}^{r}\sum_{k=1}^{n-1}\sum_{l=1}^{n-1}\frac{1}{\epsilon_{nkl}^2}\beta_{nkj}^2\beta_{nlj}^2 + \frac{3}{4}\sum_{i=1}^{n-1}z_i^4\sum_{j=1}^{r}\sum_{k=1}^{i-1}\sum_{l=1}^{i-1}\frac{1}{\epsilon_{ikl}^2}\beta_{ikj}^2\beta_{ilj}^2$$

$$+\frac{3r}{4}\sum_{i=1}^{n-1}z_i^4\sum_{k=i+1}^{n}\sum_{l=1}^{k-1}\epsilon_{kil}^2$$

$$= z_n^3\left(u - \sum_{l=1}^{n-1}\frac{\partial\alpha_{n-1}}{\partial x_l}x_{l+1} - \frac{1}{2}\sum_{p,q=1}^{n-1}\frac{\partial^2\alpha_{n-1}}{\partial x_p\partial x_q}\varphi_p^T\varphi_q + \frac{1}{4\epsilon_{n-1}^4}z_n\right.$$

$$\left.+\frac{3}{2}z_n\beta_{nn}^T\beta_{nn} + 3\beta_{nn}^T\sum_{k=1}^{n-1}z_k\beta_{nk} + \frac{3}{4}z_n\sum_{j=1}^{r}\sum_{k=1}^{n-1}\sum_{l=1}^{n-1}\frac{1}{\epsilon_{nkl}^2}\beta_{nkj}^2\beta_{nlj}^2\right)$$

$$+z_1^3\left(\alpha_1 + \frac{3}{4}\epsilon_1^{\frac{4}{3}}z_1 + \frac{3}{2}z_1\beta_{11}^T\beta_{11} + \frac{3r}{4}z_1\sum_{k=2}^{n}\sum_{l=1}^{k-1}\epsilon_{k1l}^2\right)$$

$$+\sum_{i=2}^{n-1}z_i^3\left(\alpha_i - \sum_{l=1}^{i-1}\frac{\partial\alpha_{i-1}}{\partial x_l}x_{l+1} - \frac{1}{2}\sum_{p,q=1}^{i-1}\frac{\partial^2\alpha_{i-1}}{\partial x_p\partial x_q}\varphi_p^T\varphi_q + \frac{3}{4}\epsilon_i^{\frac{4}{3}}z_i\right.$$

3.2 STABILIZATION IN PROBABILITY VIA BACKSTEPPING

$$+\frac{1}{4\epsilon_{i-1}^4}z_i + \frac{3}{2}z_i\beta_{ii}^T\beta_{ii} + 3\beta_{ii}^T\sum_{k=1}^{i-1}z_k\beta_{ik} + \frac{3}{4}z_i\sum_{j=1}^{r}\sum_{k=1}^{i-1}\sum_{l=1}^{i-1}\frac{1}{\epsilon_{ikl}^2}\beta_{ikj}^2\beta_{ilj}^2$$

$$+\frac{3r}{4}z_i\sum_{k=i+1}^{n}\sum_{l=1}^{k-1}\epsilon_{kil}^2 \Bigg). \tag{3.50}$$

At this point, all the terms can be cancelled by u and α_i. If we choose them as

$$\alpha_1 = -c_1 z_1 - \frac{3}{4}\epsilon_1^{\frac{4}{3}} z_1 - \frac{3}{2}z_1\beta_{11}^T\beta_{11} - \frac{3r}{4}z_1\sum_{k=2}^{n}\sum_{l=1}^{k-1}\epsilon_{k1l}^2 \tag{3.51}$$

$$\alpha_i = -c_i z_i + \sum_{l=1}^{i-1}\frac{\partial\alpha_{i-1}}{\partial x_l}x_{l+1} + \frac{1}{2}\sum_{p,q=1}^{i-1}\frac{\partial^2\alpha_{i-1}}{\partial x_p \partial x_q}\varphi_p^T\varphi_q$$

$$-\frac{3}{4}\epsilon_i^{\frac{4}{3}} z_i - \frac{1}{4\epsilon_{i-1}^4}z_i - \frac{3}{2}z_i\beta_{ii}^T\beta_{ii} - 3\beta_{ii}^T\sum_{k=1}^{i-1}z_k\beta_{ik}$$

$$-\frac{3}{4}z_i\sum_{j=1}^{r}\sum_{k=1}^{i-1}\sum_{l=1}^{i-1}\frac{1}{\epsilon_{ikl}^2}\beta_{ikj}^2\beta_{ilj}^2 - \frac{3r}{4}z_i\sum_{k=i+1}^{n}\sum_{l=1}^{k-1}\epsilon_{kil}^2 \tag{3.52}$$

$$u = -c_n z_n + \sum_{l=1}^{n-1}\frac{\partial\alpha_{n-1}}{\partial x_l}x_{l+1} + \frac{1}{2}\sum_{p,q=1}^{n-1}\frac{\partial^2\alpha_{n-1}}{\partial x_p \partial x_q}\varphi_p^T\varphi_q - \frac{1}{4\epsilon_{n-1}^4}z_n - \frac{3}{2}z_n\beta_{nn}^T\beta_{nn}$$

$$-3\beta_{nn}^T\sum_{k=1}^{n-1}z_k\beta_{nk} - \frac{3}{4}z_n\sum_{j=1}^{r}\sum_{k=1}^{n-1}\sum_{l=1}^{n-1}\frac{1}{\epsilon_{nkl}^2}\beta_{nkj}^2\beta_{nlj}^2 \tag{3.53}$$

where $c_i > 0$, then the infinitesimal generator of the system (3.38) is negative definite:

$$\mathcal{L}V \leq -\sum_{i=1}^{n}c_i z_i^4. \tag{3.54}$$

We want to derive a systematic design procedure which can be coded using symbolic software. By collecting expressions (3.34), (3.35), (3.45), (3.51), (3.52), and (3.53), and by performing necessary rearrangements, we get the algorithm given in Table 3.1. The only input the user needs to provide at the start of the algorithm are functions φ_{ik} for the factorization of the system nonlinearities, $\varphi_i = \sum_{k=1}^{i} x_k \varphi_{ik}$.

With (3.54), we have the following stability result for the design procedure in Table 3.1.

Theorem 3.4 *The equilibrium at the origin of the closed-loop stochastic system (3.36), (3.37), (3.62) is globally asymptotically stable in probability. Furthermore, the following estimate of the 4th-moment exponential stability is guaranteed:*

$$E\left\{|z(t)|_4^4\right\} \leq |z(0)|_4^4 e^{-4ct}, \tag{3.63}$$

Table 3.1: Backstepping Design for System (3.28), (3.29)

$$z_i = x_i - \alpha_{i-1}, \qquad i = 1,\ldots,n \tag{3.55}$$

$$\psi_{ik} = \sum_{l=k}^{i} \alpha_{l-1,k}\varphi_{il}, \qquad k=1,\ldots,i \tag{3.56}$$

$$\beta_{ik} = \psi_{ik} - \sum_{l=k}^{i-1} \frac{\partial \alpha_{i-1}}{\partial x_l}\psi_{lk}, \qquad k=1,\ldots,i \tag{3.57}$$

$$\alpha_{ik} = \frac{1}{2}\sum_{p=k}^{i-1}\psi_{pk}^{\mathrm{T}}\sum_{q=1}^{i-1}\frac{\partial^2 \alpha_{i-1}}{\partial x_p \partial x_q}\sum_{j=1}^{q}z_j\psi_{qj}$$
$$+ \sum_{l=k-1}^{i-1}\frac{\partial \alpha_{i-1}}{\partial x_l}\alpha_{lk} - 3\beta_{ii}^{\mathrm{T}}\beta_{ik}, \qquad k=1,\ldots,i-1 \tag{3.58}$$

$$\alpha_{ii} = -\left(c_i + \frac{3}{4}\epsilon_i^{\frac{4}{3}} + \frac{1}{4\epsilon_{i-1}^4} + \frac{3}{2}\beta_{ii}^{\mathrm{T}}\beta_{ii} + \frac{3r}{4}\sum_{k=i+1}^{n}\sum_{l=1}^{k-1}\epsilon_{kil}^2 \right.$$
$$\left. + \frac{3}{4}\sum_{j=1}^{r}\sum_{k=1}^{i-1}\sum_{l=1}^{i-1}\frac{1}{\epsilon_{ikl}^2}\beta_{ikj}^2\beta_{ilj}^2\right) + \frac{\partial \alpha_{i-1}}{\partial x_{i-1}} \tag{3.59}$$

$$\alpha_{i,i+1} = 1 \tag{3.60}$$

$$\alpha_i = \sum_{k=1}^{i} z_k \alpha_{ik} \tag{3.61}$$

Control law:
$$u = \alpha_n \tag{3.62}$$

where $|z|_4 = \left(\sum_i z_i^4\right)^{1/4}$ and $c = \min_i c_i$.

Proof. The first part of the theorem follows from (3.54) by Theorem 3.3. As to the second part, we just give the main point of the proof. According to [59, Lemma 3.1, p. 82], we have

$$\frac{d}{dt}\left\{E\left[\frac{1}{4}\sum z_i^4\right]\right\} = E\{\mathcal{L}V\}, \tag{3.64}$$

which yields the differential inequality

$$\frac{d}{dt}E\{\sum z_i^4\} \leq -4cE\{\sum z_i^4\}, \tag{3.65}$$

and gives (3.63). □

3.3 Stochastic Control Lyapunov Functions

Consider the system which, in addition to the noise input w, has a control input u:

$$dx = f(x)dt + g_1(x)dw + g_2(x)udt, \qquad (3.66)$$

where $f(0) = 0$, $g_1(0) = 0$ and $u \in \mathbb{R}^m$. We say that the system is *globally asymptotically stabilizable in probability* if there exists a control law $u = \alpha(x)$ continuous everywhere, with $\alpha(0) = 0$, such that the equilibrium $x = 0$ of the closed-loop system is globally asymptotically stable in probability.

Definition 3.5 *A smooth positive definite radially unbounded function $V : \mathbb{R}^n \to \mathbb{R}_+$ is called a **stochastic control Lyapunov function** (sclf) for system (3.66) if it satisfies*

$$\inf_{u \in \mathbb{R}^m} \left\{ L_f V + \frac{1}{2} \mathrm{Tr}\left\{ g_1^\mathrm{T} \frac{\partial^2 V}{\partial x^2} g_1 \right\} + L_{g_2} V u \right\} < 0, \qquad \forall x \neq 0. \qquad (3.67)$$

Lemma 3.6 *A positive definite radially unbounded function $V(x)$ is an sclf if and only if for all $x \neq 0$,*

$$L_{g_2} V = 0 \Rightarrow L_f V + \frac{1}{2} \mathrm{Tr}\left\{ g_1^\mathrm{T} \frac{\partial^2 V}{\partial x^2} g_1 \right\} < 0. \qquad (3.68)$$

The existence of an sclf guarantees global asymptotic stabilizability in probability, as shown in the following theorem.

Theorem 3.7 *The system (3.66) is globally asymptotically stabilizable in probability if there exists an sclf with small control property.*

Proof. Consider the Sontag-type formula $u = \alpha_s(x)$,

$$\alpha_s(x) = \begin{cases} -\dfrac{\omega + \sqrt{\omega^2 + \left(L_{g_2} V (L_{g_2} V)^\mathrm{T}\right)^2}}{L_{g_2} V (L_{g_2} V)^\mathrm{T}} (L_{g_2} V)^\mathrm{T}, & L_{g_2} V \neq 0 \\ 0, & L_{g_2} V = 0 \end{cases} \qquad (3.69)$$

where

$$\omega = L_f V + \frac{1}{2} \mathrm{Tr}\left\{ g_1^\mathrm{T} \frac{\partial^2 V}{\partial x^2} g_1 \right\}. \qquad (3.70)$$

According to Lemmas 3.6, 1.11 and 1.14, this formula is smooth away from the origin and continuous at the origin. Substituting (3.69) into (3.3) we get

$$\mathcal{L}V = -\sqrt{\omega^2 + (L_{g_2} V (L_{g_2} V)^\mathrm{T})^2} \qquad (3.71)$$

which is positive definite by Lemma 3.6. Thus by Theorem 3.3, global asymptotic stability in probability is assured. □

3.4 Inverse Optimal Stabilization in Probability

Definition 3.8 *The problem of **inverse optimal stabilization in probability** for system (3.66) is solvable if there exist a class $\mathcal{K}_\infty$ function γ_2 whose derivative γ_2' is also a class $\mathcal{K}_\infty$ function, a matrix-valued function $R_2(x)$ such that $R_2(x) = R_2(x)^{\mathrm{T}} > 0$ for all x, a positive definite function $l(x)$, and a feedback control law $u = \alpha(x)$ continuous everywhere with $\alpha(0) = 0$, which guarantees global asymptotic stability in probability of the equilibrium $x = 0$ and minimizes the cost functional*

$$J(u) = E\left\{\int_0^\infty \left[l(x) + \gamma_2\left(\left|R_2(x)^{1/2}u\right|\right)\right] d\tau\right\}. \quad (3.72)$$

In the next theorem we extensively use the Legendre-Fenchel transform given in Appendix A as follows: For a class $\mathcal{K}_\infty$ function γ, whose derivative γ' is also a class $\mathcal{K}_\infty$ function, $\ell\gamma$ denotes

$$\ell\gamma(r) = \int_0^r (\gamma')^{-1}(s)ds. \quad (3.73)$$

Theorem 3.9 *Consider the control law*

$$u = \alpha(x) = -R_2^{-1}(L_{g_2}V)^{\mathrm{T}} \frac{\ell\gamma_2\left(\left|L_{g_2}VR_2^{-1/2}\right|\right)}{\left|L_{g_2}VR_2^{-1/2}\right|^2}, \quad (3.74)$$

where $V(x)$ is a Lyapunov function candidate, γ_2 is a class $\mathcal{K}_\infty$ function whose derivative is also a class $\mathcal{K}_\infty$ function, and $R_2(x)$ is a matrix-valued function such that $R_2(x) = R_2(x)^{\mathrm{T}} > 0$. If the control law (3.74) achieves global asymptotic stability in probability for the system (3.66) with respect to $V(x)$, then the control law

$$u^* = \alpha^*(x) = -\frac{\beta}{2}R_2^{-1}(L_{g_2}V)^{\mathrm{T}} \frac{(\gamma_2')^{-1}\left(\left|L_{g_2}VR_2^{-1/2}\right|\right)}{\left|L_{g_2}VR_2^{-1/2}\right|}, \quad \beta \geq 2 \quad (3.75)$$

solves the problem of inverse optimal stabilization in probability for the system (3.66) by minimizing the cost functional

$$J(u) = E\left\{\int_0^\infty \left[l(x) + \beta^2\gamma_2\left(\frac{2}{\beta}\left|R_2^{1/2}u\right|\right)\right] d\tau\right\}, \quad (3.76)$$

where

$$\begin{aligned}l(x) &= 2\beta\left[\ell\gamma_2\left(\left|L_{g_2}VR_2^{-1/2}\right|\right) - L_fV - \frac{1}{2}\mathrm{Tr}\left\{g_1^{\mathrm{T}}\frac{\partial^2 V}{\partial x^2}g_1\right\}\right] \\ &\quad + \beta(\beta-2)\ell\gamma_2\left(\left|L_{g_2}VR_2^{-1/2}\right|\right).\end{aligned} \quad (3.77)$$

3.4 INVERSE OPTIMAL STABILIZATION IN PROBABILITY

Proof. Since the control law (3.74) globally asymptotically stabilizes the system in probability, then there exists a continuous positive definite function $W : \mathbb{R}^n \to \mathbb{R}_+$ such that

$$\begin{aligned}
\mathcal{L}V \mid_{(3.74)} &= L_f V + \frac{1}{2}\text{Tr}\left\{g_1^T \frac{\partial^2 V}{\partial x^2} g_1\right\} + L_{g_2} V \alpha(x) \\
&= -\ell\gamma_2\left(\left|L_{g_2} V R_2^{-1/2}\right|\right) + L_f V + \frac{1}{2}\text{Tr}\left\{g_1^T \frac{\partial^2 V}{\partial x^2} g_1\right\} \\
&\leq -W(x). \quad (3.78)
\end{aligned}$$

Then we have

$$l(x) \geq 2\beta W(x) + \beta(\beta - 2)\ell\gamma_2\left(\left|L_{g_2} V R_2^{-1/2}\right|\right). \quad (3.79)$$

Since $W(x)$ is positive definite, $\beta \geq 2$, and $\ell\gamma_2$ is a class $\mathcal{K}_\infty$ function (Lemma A.1), $l(x)$ is positive definite. Therefore $J(u)$ is a meaningful cost functional.

Before we engage into proving that the control law (3.75) minimizes (3.76), we first show that it is stabilizing. With Lemma A.1 we get

$$\begin{aligned}
\mathcal{L}V \mid_{(3.75)} &= L_f V + \frac{1}{2}\text{Tr}\left\{g_1^T \frac{\partial^2 V}{\partial x^2} g_1\right\} - \frac{\beta}{2}\left|L_{g_2} V R_2^{-1/2}\right|(\gamma_2')^{-1}\left(\left|L_{g_2} V R_2^{-1/2}\right|\right) \\
&= L_f V + \frac{1}{2}\text{Tr}\left\{g_1^T \frac{\partial^2 V}{\partial x^2} g_1\right\} - \frac{\beta}{2}\Big[\ell\gamma_2\left(\left|L_{g_2} V R_2^{-1/2}\right|\right) \\
&\quad + \gamma_2\left((\gamma_2')^{-1}\left(\left|L_{g_2} V R_2^{-1/2}\right|\right)\right)\Big] \\
&\leq \mathcal{L}V \mid_{(3.74)} < 0, \quad \forall x \neq 0, \quad (3.80)
\end{aligned}$$

which proves that (3.75) achieves global asymptotic stability in probability and, in particular, that $x(t) \to 0$ with probability 1.

Now we prove optimality. Recalling that the Itô differential of V is

$$dV = \mathcal{L}V(x)dt + \frac{\partial V}{\partial x} g_1(x) dw, \quad (3.81)$$

according to the property of Itô's integral [86, Theorem 3.9], we get

$$E\left\{V(0) - V(t) + \int_0^t \mathcal{L}V(x(\tau))d\tau\right\} = 0. \quad (3.82)$$

Then substituting $l(x)$ into $J(u)$, we have

$$\begin{aligned}
J(u) &= E\left\{\int_0^\infty \left[l(x) + \beta^2 \gamma_2\left(\frac{2}{\beta}\left|R_2^{1/2} u\right|\right)\right]d\tau\right\} \\
&= 2\beta E\{V(x(0))\} - 2\beta \lim_{t\to\infty} E\{V(x(t))\} \\
&\quad + E\left\{\int_0^\infty \left[2\beta\mathcal{L}V \mid_{(3.66)} + l(x) + \beta^2 \gamma_2\left(\frac{2}{\beta}\left|R_2^{1/2} u\right|\right)\right]d\tau\right\}
\end{aligned}$$

$$= 2\beta E\left\{V(x(0))\right\} - 2\beta \lim_{t\to\infty} E\left\{V(x(t))\right\}$$
$$+ E\left\{\int_0^\infty \left[\beta^2\gamma_2\left(\frac{2}{\beta}|R_2^{1/2}u|\right) + \beta^2\ell\gamma_2\left(\left|L_{g_2}VR_2^{-1/2}\right|\right)\right.\right.$$
$$\left.\left. + 2\beta L_{g_2}Vu\right]d\tau\right\}. \tag{3.83}$$

Now we note that

$$\gamma_2'\left(\frac{2}{\beta}|R_2^{1/2}u^*|\right) = \left|L_{g_2}VR_2^{-1/2}\right|, \tag{3.84}$$

which yields

$$J(u) = E\left\{\int_0^\infty \left[\beta^2\gamma_2\left(\left|\frac{2}{\beta}R_2^{1/2}u\right|\right) + \beta^2\ell\gamma_2\left(\gamma_2'\left(\left|\frac{2}{\beta}R_2^{1/2}u^*\right|\right)\right)\right.\right.$$
$$\left.\left. - 2\beta\gamma_2'\left(\left|\frac{2}{\beta}R_2^{1/2}u^*\right|\right)\frac{\left(\frac{2}{\beta}R_2^{1/2}u^*\right)^T}{\left|\frac{2}{\beta}R_2^{1/2}u^*\right|}R_2^{1/2}u\right]d\tau\right\}$$
$$+ 2\beta E\left\{V(x(0))\right\} - 2\beta \lim_{t\to\infty} E\left\{V(x(t))\right\}. \tag{3.85}$$

With the general Young's inequality (Lemma A.2), we obtain

$$J(u) \geq 2\beta E\left\{V(x(0))\right\} + E\left\{\int_0^\infty \left[\beta^2\gamma_2\left(\left|\frac{2}{\beta}R_2^{1/2}u\right|\right)\right.\right.$$
$$\left.+ \beta^2\ell\gamma_2\left(\gamma_2'\left(\left|\frac{2}{\beta}R_2^{1/2}u^*\right|\right)\right) - \beta^2\gamma_2\left(\left|\frac{2}{\beta}R_2^{1/2}u\right|\right)\right.$$
$$\left.\left. - \beta^2\ell\gamma_2\left(\gamma_2'\left(\left|\frac{2}{\beta}R_2^{1/2}u^*\right|\right)\right)\right]d\tau\right\} - 2\beta \lim_{t\to\infty} E\left\{V(x(t))\right\}$$
$$= 2\beta E\left\{V(x(0))\right\} - 2\beta \lim_{t\to\infty} E\left\{V(x(t))\right\}, \tag{3.86}$$

where the equality holds if and only if

$$\gamma_2'\left(\left|\frac{2}{\beta}R_2^{1/2}u^*\right|\right)\frac{\left(\frac{2}{\beta}R_2^{1/2}u^*\right)^T}{\left|\frac{2}{\beta}R_2^{1/2}u^*\right|} = \gamma_2'\left(\left|\frac{2}{\beta}R_2^{1/2}u\right|\right)\frac{\left(\frac{2}{\beta}R_2^{1/2}u\right)^T}{\left|\frac{2}{\beta}R_2^{1/2}u\right|}, \tag{3.87}$$

that is, $u = u^*$. Since $u = u^*$ is stabilizing in probability, $\lim_{t\to\infty} E\left\{V(x(t))\right\} = 0$, and thus

$$\arg\min_u J(u) = u^* \tag{3.88}$$
$$\min_u J(u) = 2\beta E\left\{V(x(0))\right\}. \tag{3.89}$$

To satisfy the requirements of Definition 3.8, it only remains to prove that $\alpha^*(x)$ is continuous and $\alpha^*(0) = 0$. Because g_2, R_2 and $\partial V/\partial x$ are continuous

3.4 INVERSE OPTIMAL STABILIZATION IN PROBABILITY

functions, and $(\gamma_2')^{-1}$ is a class $\mathcal{K}_\infty$ function, $\alpha^*(x)$ is continuous away from $L_{g_2}VR_2^{-1/2} = 0$. If $\left|L_{g_2}VR_2^{-1/2}\right| \to 0$, continuity can be inferred from the fact that

$$\lim_{\left|L_{g_2}VR_2^{-1/2}\right| \to 0} |\alpha^*(x)|$$

$$= \lim_{\left|L_{g_2}VR_2^{-1/2}\right| \to 0} \left\{ \frac{\beta}{2} \left|R_2^{-1/2}\right| \left|L_{g_2}VR_2^{-1/2}\right| \frac{(\gamma_2')^{-1}\left(\left|L_{g_2}VR_2^{-1/2}\right|\right)}{\left|L_{g_2}VR_2^{-1/2}\right|} \right\}$$

$$= \lim_{\left|L_{g_2}VR_2^{-1/2}\right| \to 0} \left\{ \frac{\beta}{2} \left|R_2^{-1/2}\right| (\gamma_2')^{-1}\left(\left|L_{g_2}VR_2^{-1/2}\right|\right) \right\}$$

$$= 0. \tag{3.90}$$

Since $\dfrac{\partial V}{\partial x}(0) = 0$, $L_{g_2}V(0) = 0$ and we have $\alpha^*(0) = 0$. □

Remark 3.10 Even though not explicit in the proof of Theorem 3.9, $V(x)$ solves the following family of *Hamilton-Jacobi-Bellman* equations parameterized by $\beta \in [2, \infty)$:

$$L_f V + \frac{1}{2}\text{Tr}\left\{g_1^\text{T} \frac{\partial^2 V}{\partial x^2} g_1\right\} - \frac{\beta}{2}\ell\gamma_2\left(\left|L_{g_2}VR_2^{-1/2}\right|\right) + \frac{l(x)}{2\beta} = 0. \tag{3.91}$$

□

In the next theorem we design controllers which are inverse optimal in the sense of Definition 3.8.

Theorem 3.11 *If the system (3.66) has an sclf with a small control property, then the problem of inverse optimal stabilization in probability is solvable.*

Proof. Let $\gamma_2(r) = \dfrac{1}{4}r^2$, $\beta = 2$, and

$$R_2(x) = I \begin{cases} -\dfrac{2L_{g_2}V\,(L_{g_2}V)^\text{T}}{\omega + \sqrt{\omega^2 + \left(L_{g_2}V\,(L_{g_2}V)^\text{T}\right)^2}}, & L_{g_2}V \neq 0 \\ \text{any positive number}, & L_{g_2}V = 0. \end{cases} \tag{3.92}$$

Then, with $(\gamma_2')^{-1}(r) = 2r$, the optimal control candidate (3.75) is given by the formula (3.69), $u^* = \alpha_s(x)$. To prove that $\alpha_s(x)$ is inverse optimal, we prove that the control law (3.74) is stabilizing. Since $\ell\gamma_2(r) = r^2$, (3.74) becomes $u = \dfrac{1}{2}\alpha_s(x)$. Then the infinitesimal generator of the system (3.66) with the control law $u = \dfrac{1}{2}\alpha_s(x)$ is

$$\mathcal{L}V = -\frac{1}{2}\left[-\omega + \sqrt{\omega^2 + \left(L_{g_2}V\,(L_{g_2}V)^\text{T}\right)^2}\right], \tag{3.93}$$

which is negative definite by Lemma 3.6. □

Remark 3.12 In the stochastic case we do not try to make $l(x)$ radially unbounded and $R_2(x) = I$ (and γ_2 quadratic) with the objective of achieving stability margins as in Section 2.4. The obstacle to achieving this is the Hessian term $\partial^2 V / \partial x^2$ which prevents simple modifications of the Lyapunov function as in (2.89). □

3.5 Inverse Optimal Backstepping

The controller given in Table 3.1 is stabilizing but is not inverse optimal because it is not of the form (3.75). In this section we will redesign the stabilizing functions $\alpha_i(x)$ to get an inverse optimal control law.

To design a control law in the form (3.75), we first note from $V = \sum_{i=1}^{n} \frac{1}{4} z_i^4$ that for the system (3.28), (3.29) we have $L_{g_2} V = z_n^3$ and, since u is a scalar input, $R_2(x)$ is sought as a scalar positive function. The following lemma is instrumental in motivating the inverse optimal design.

Lemma 3.13 *If there exists a continuous positive function $M(x)$ such that the control law*

$$u = \alpha(x) = -M(x) z_n \tag{3.94}$$

globally asymptotically stabilizes system (3.28), (3.29) in probability with respect to the Lyapunov function $V = \sum_{i=1}^{n} \frac{1}{4} z_i^4$, then the control law

$$u^* = \alpha^*(x) = \frac{2}{3} \beta \alpha(x), \qquad \beta \geq 2 \tag{3.95}$$

solves the problem of inverse optimal stabilization in probability.

Proof. Let

$$\gamma_2(r) = \frac{1}{4} r^4 \tag{3.96}$$

$$R_2 = \left(\frac{4}{3} M\right)^{-3/2}. \tag{3.97}$$

Then the control laws (3.74) and (3.75) are given, respectively, as

$$u = \alpha(x) = -\frac{3}{4} R_2^{-2/3} z_n, \tag{3.98}$$

$$u^* = \alpha^*(x) = -\frac{\beta}{2} R_2^{-2/3} z_n. \tag{3.99}$$

By dividing the last two expressions, we get (3.95), and the lemma follows by Theorem 3.9. □

3.5 INVERSE OPTIMAL BACKSTEPPING

From the lemma it follows that we should seek a stabilizing control law in the form of (3.94). If we consider carefully the parenthesis including u in (3.50), every term except the second, the third, and the sixth has z_n as a factor. We now deal with the three terms one at a time. The second term yields

$$-z_n^3 \sum_{l=1}^{n-1} \frac{\partial \alpha_{n-1}}{\partial x_l} x_{l+1}$$

$$= -z_n^3 \sum_{l=1}^{n-1} \frac{\partial \alpha_{n-1}}{\partial x_l} \left(z_{l+1} + \sum_{k=1}^{l} z_k \alpha_{lk} \right)$$

$$= -z_n^4 \frac{\partial \alpha_{n-1}}{\partial x_{n-1}} - \sum_{l=1}^{n-2} z_n^3 \frac{\partial \alpha_{n-1}}{\partial x_l} z_{l+1} - \sum_{l=1}^{n-1} z_n^3 \frac{\partial \alpha_{n-1}}{\partial x_l} \sum_{k=1}^{l} z_k \alpha_{lk}$$

$$\leq z_n^4 \left(\frac{1}{2} + \frac{1}{2} \left(\frac{\partial \alpha_{n-1}}{\partial x_{n-1}} \right)^2 \right) + \sum_{l=1}^{n-2} \left[\frac{3}{4} z_n^4 \left(\delta_l \frac{\partial \alpha_{n-1}}{\partial x_l} \right)^{\frac{4}{3}} + \frac{1}{4\delta_l^4} z_{l+1}^4 \right]$$

$$+ \sum_{l=1}^{n-1} \sum_{k=1}^{l} \left[\frac{3}{4} z_n^4 \left(\delta_{lk} \frac{\partial \alpha_{n-1}}{\partial x_l} \alpha_{lk} \right)^{\frac{4}{3}} + \frac{1}{4\delta_{lk}^4} z_k^4 \right]$$

$$= z_n^4 \left[\left(\frac{1}{2} + \frac{1}{2} \left(\frac{\partial \alpha_{n-1}}{\partial x_{n-1}} \right)^2 \right) + \frac{3}{4} \sum_{l=1}^{n-2} \left(\delta_l \frac{\partial \alpha_{n-1}}{\partial x_l} \right)^{\frac{4}{3}} + \frac{3}{4} \sum_{l=1}^{n-1} \sum_{k=1}^{l} \left(\delta_{lk} \frac{\partial \alpha_{n-1}}{\partial x_l} \alpha_{lk} \right)^{\frac{4}{3}} \right]$$

$$+ \sum_{l=1}^{n-1} \sum_{k=1}^{l} \frac{1}{4\delta_{lk}^4} z_k^4 + \sum_{l=2}^{n-1} \frac{1}{4\delta_{l-1}^4} z_l^4$$

$$= z_n^4 \left[\left(\frac{1}{2} + \frac{1}{2} \left(\frac{\partial \alpha_{n-1}}{\partial x_{n-1}} \right)^2 \right) + \frac{3}{4} \sum_{l=1}^{n-2} \left(\delta_l \frac{\partial \alpha_{n-1}}{\partial x_l} \right)^{\frac{4}{3}} + \frac{3}{4} \sum_{l=1}^{n-1} \sum_{k=1}^{l} \left(\delta_{lk} \frac{\partial \alpha_{n-1}}{\partial x_l} \alpha_{lk} \right)^{\frac{4}{3}} \right]$$

$$+ \sum_{k=1}^{n-1} z_k^4 \sum_{l=k}^{n-1} \frac{1}{4\delta_{lk}^4} + \sum_{l=2}^{n-1} \frac{1}{4\delta_{l-1}^4} z_l^4, \qquad (3.100)$$

where the inequality is obtained by applying Young's inequality,

$$-z_n^3 \frac{\partial \alpha_{n-1}}{\partial x_l} z_{l+1} \leq \frac{3}{4} z_n^4 \left(\delta_l \frac{\partial \alpha_{n-1}}{\partial x_l} \right)^{\frac{4}{3}} + \frac{1}{4\delta_l^4} z_{l+1}^4 \qquad (3.101)$$

$$z_n^3 \frac{\partial \alpha_{n-1}}{\partial x_l} z_k \alpha_{lk} \leq \frac{3}{4} z_n^4 \left(\delta_{lk} \frac{\partial \alpha_{n-1}}{\partial x_l} \alpha_{lk} \right)^{\frac{4}{3}} + \frac{1}{4\delta_{lk}^4} z_k^4 \qquad (3.102)$$

$$-\frac{\partial \alpha_{n-1}}{\partial x_{n-1}} \leq \frac{1}{2} + \frac{1}{2} \left(\frac{\partial \alpha_{n-1}}{\partial x_{n-1}} \right)^2, \qquad (3.103)$$

and the last equality comes from changing the summation order in the second term. As to the third and the sixth terms, in the first parenthesis in (3.50), we have

$$-\frac{1}{2}z_n^3 \sum_{p,q=1}^{n-1} \frac{\partial^2 \alpha_{n-1}}{\partial x_p \partial x_q} \varphi_p^{\mathrm{T}} \varphi_q$$

$$= -\frac{1}{2}z_n^3 \sum_{p,q=1}^{n-1} \frac{\partial^2 \alpha_{n-1}}{\partial x_p \partial x_q} \left(\sum_{k=1}^{p} z_k \psi_{pk}\right)^{\mathrm{T}} \left(\sum_{l=1}^{q} z_l \psi_{ql}\right)$$

$$= -\frac{1}{2} \sum_{p,q=1}^{n-1} \sum_{k=1}^{p} \sum_{l=1}^{q} \left(z_n^3 \frac{\partial^2 \alpha_{n-1}}{\partial x_p \partial x_q} \psi_{pk}^{\mathrm{T}} \psi_{ql}\right) z_k z_l$$

$$\leq \frac{1}{2} \sum_{p,q=1}^{n-1} \sum_{k=1}^{p} \sum_{l=1}^{q} \left[\frac{3}{4}\delta_{pqkl}^{\frac{4}{3}} z_n^4 \left(\frac{\partial^2 \alpha_{n-1}}{\partial x_p \partial x_q} \psi_{pk}^{\mathrm{T}} \psi_{ql} z_k\right)^{\frac{4}{3}} + \frac{1}{4\delta_{pqkl}^4} z_l^4\right]$$

$$\leq \frac{3}{8}z_n^4 \sum_{p,q=1}^{n-1} \sum_{k=1}^{p} \sum_{l=1}^{q} \left(\delta_{pqkl} \frac{\partial^2 \alpha_{n-1}}{\partial x_p \partial x_q} \psi_{pk}^{\mathrm{T}} \psi_{ql} z_k\right)^{\frac{4}{3}} + \frac{1}{2} \sum_{p,q=1}^{n-1} \sum_{k=1}^{p} \sum_{l=1}^{q} \frac{1}{4\delta_{pqkl}^4} z_l^4$$

$$= \frac{3}{8}z_n^4 \sum_{p,q=1}^{n-1} \sum_{k=1}^{p} \sum_{l=1}^{q} \left(\delta_{pqkl} \frac{\partial^2 \alpha_{n-1}}{\partial x_p \partial x_q} \psi_{pk}^{\mathrm{T}} \psi_{ql} z_k\right)^{\frac{4}{3}} + \frac{1}{2} \sum_{l=1}^{n-1} z_l^4 \sum_{p=1}^{n-1} \sum_{q=l}^{n-1} \sum_{k=1}^{p} \frac{1}{4\delta_{pqkl}^4}$$

(3.104)

and

$$3z_n^3 \beta_{nn}^{\mathrm{T}} \sum_{k=1}^{n-1} z_k \beta_{nk} = 3 \sum_{k=1}^{n-1} \left(z_n^3 \beta_{nn}^{\mathrm{T}} \beta_{nk}\right) z_k$$

$$\leq 3 \sum_{k=1}^{n-1} \left[\frac{3}{4} \left(\delta_k \beta_{nn}^{\mathrm{T}} \beta_{nk}\right)^{\frac{4}{3}} z_n^4 + \frac{1}{4\delta_k^4} z_k^4\right]$$

$$= \frac{9}{4} z_n^4 \sum_{k=1}^{n-1} \left(\delta_k \beta_{nn}^{\mathrm{T}} \beta_{nk}\right)^{\frac{4}{3}} + \frac{3}{4} \sum_{k=1}^{n-1} \frac{1}{\delta_k^4} z_k^4. \quad (3.105)$$

Substituting (3.100), (3.104) and (3.105) into (3.50), we have

$$\mathcal{L}V \leq z_n^3 \left\{ u + z_n \left[\left(\frac{1}{2} + \frac{1}{2}\left(\frac{\partial \alpha_{n-1}}{\partial x_{n-1}}\right)^2\right)\right.\right.$$

$$\left.+ \frac{3}{4} \sum_{l=1}^{n-2} \left(\delta_l \frac{\partial \alpha_{n-1}}{\partial x_l}\right)^{\frac{4}{3}} + \frac{3}{4} \sum_{l=1}^{n-1} \sum_{k=1}^{l} \left(\delta_{lk} \frac{\partial \alpha_{n-1}}{\partial x_l} \alpha_{lk}\right)^{\frac{4}{3}}\right]$$

$$+ \frac{3}{8} z_n \sum_{p,q=1}^{n-1} \sum_{k=1}^{p} \sum_{l=1}^{q} \left(\delta_{pqkl} \frac{\partial^2 \alpha_{n-1}}{\partial x_p \partial x_q} \psi_{pk}^{\mathrm{T}} \psi_{ql} z_k\right)^{\frac{4}{3}} + \frac{1}{4\epsilon_{n-1}^4} z_n + \frac{3}{2} z_n \beta_{nn}^{\mathrm{T}} \beta_{nn}$$

$$+ \frac{9}{4} z_n \sum_{k=1}^{n-1} \left(\delta_k \beta_{nn}^{\mathrm{T}} \beta_{nk}\right)^{\frac{4}{3}} + \frac{3}{4} z_n \sum_{j=1}^{r} \sum_{k=1}^{n-1} \sum_{l=1}^{n-1} \frac{1}{\epsilon_{nkl}^2} \beta_{nkj}^2 \beta_{nlj}^2 \right\}$$

$$+ z_1^3 \left(\alpha_1 + z_1 \sum_{l=1}^{n-1} \frac{1}{4\delta_{l1}^4} + \frac{3}{4}\epsilon_1^{\frac{4}{3}} z_1 + \frac{1}{2} z_1 \sum_{p=1}^{n-1}\sum_{q=1}^{n-1}\sum_{k=1}^{p} \frac{1}{4\delta_{pqk1}^4} + \frac{3}{4}\frac{1}{\delta_1^4} z_1\right.$$

3.5 INVERSE OPTIMAL BACKSTEPPING

$$+\frac{3}{2}z_1\beta_{11}^T\beta_{11} + \frac{3r}{4}z_1\sum_{k=2}^{n}\sum_{l=1}^{k-1}\epsilon_{k1l}^2\Bigg)$$

$$+\sum_{i=2}^{n-1}z_i^3\Bigg(\alpha_i + z_i\sum_{l=i}^{n-1}\frac{1}{4\delta_{li}^4} + \frac{1}{4\delta_{i-1}^4}z_i + \frac{1}{2}z_i\sum_{p=1}^{n-1}\sum_{q=i}^{n-1}\sum_{k=1}^{p}\frac{1}{4\delta_{pqki}^4} + \frac{3}{4}\frac{1}{\delta_i^4}z_i$$

$$-\sum_{l=1}^{i-1}\frac{\partial\alpha_{i-1}}{\partial x_l}x_{l+1} - \frac{1}{2}\sum_{p,q=1}^{i-1}\frac{\partial^2\alpha_{i-1}}{\partial x_p\partial x_q}\varphi_p^T\varphi_q + \frac{3}{4}\epsilon_i^{\frac{4}{3}}z_i + \frac{1}{4\epsilon_{i-1}^4}z_i + \frac{3}{2}z_i\beta_{ii}^T\beta_{ii}$$

$$+3\beta_{ii}^T\sum_{k=1}^{i-1}z_k\beta_{ik} + \frac{3}{4}z_i\sum_{j=1}^{r}\sum_{k=1}^{i-1}\sum_{l=1}^{i-1}\frac{1}{\epsilon_{ikl}^2}\beta_{ikj}^2\beta_{ilj}^2$$

$$+\frac{3r}{4}z_i\sum_{k=i+1}^{n}\sum_{l=1}^{k-1}\epsilon_{kil}^2\Bigg). \tag{3.106}$$

If δ_{li}, δ_{pqil}, and δ_i are chosen as

$$\sum_{l=1}^{n-1}\frac{1}{4\delta_{l1}^4} + \frac{1}{2}\sum_{p=1}^{n-1}\sum_{q=1}^{n-1}\sum_{k=1}^{p}\frac{1}{4\delta_{pqk1}^4} - \frac{3}{4\delta_1^4} = \frac{c_1}{2} \tag{3.107}$$

$$\sum_{l=i}^{n-1}\frac{1}{4\delta_{li}^4} - \frac{1}{4\delta_{i-1}^4} - \frac{1}{2}\sum_{p=1}^{n-1}\sum_{q=i}^{n-1}\sum_{k=1}^{p}\frac{1}{4\delta_{pqki}^4} - \frac{3}{4\delta_i^4} = \frac{c_i}{2}, \tag{3.108}$$

where c_1 and c_i are those in (3.51) and (3.52), and

$$u = -M(x)z_n \tag{3.109}$$

$$M(x) = c_n + \frac{1}{4\epsilon_{n-1}^4} + \frac{3}{2}\beta_{nn}^T\beta_{nn} + \frac{3}{4}\sum_{j=1}^{r}\sum_{k=1}^{n-1}\sum_{l=1}^{n-1}\frac{1}{\epsilon_{nkl}^2}\beta_{nkj}^2\beta_{nlj}^2$$

$$+\left(\frac{1}{2} + \frac{1}{2}\left(\frac{\partial\alpha_{n-1}}{\partial x_{n-1}}\right)^2\right) + \frac{3}{4}\sum_{l=1}^{n-2}\left(\delta_l\frac{\partial\alpha_{n-1}}{\partial x_l}\right)^{\frac{4}{3}}$$

$$+\frac{3}{4}\sum_{l=1}^{n-1}\sum_{k=1}^{l}\left(\delta_{lk}\frac{\partial\alpha_{n-1}}{\partial x_l}\alpha_{lk}\right)^{\frac{4}{3}} + \frac{3}{8}\sum_{p,q=1}^{n-1}\sum_{k=1}^{p}\sum_{l=1}^{q}\left(\delta_{pqkl}\frac{\partial^2\alpha_{n-1}}{\partial x_p\partial x_q}\psi_{pk}^T\psi_{ql}z_k\right)^{\frac{4}{3}}$$

$$+\frac{9}{4}\sum_{k=1}^{n-1}\left(\delta_k\beta_{nn}^T\beta_{nk}\right)^{\frac{4}{3}} \tag{3.110}$$

where $c_i > 0$, $i = 1, \cdots, n$, and $M(x)$ is a positive function, with (3.107)–(3.110), we get

$$\mathcal{L}V \leq -\frac{1}{2}\sum_{i=1}^{n}c_iz_i^4. \tag{3.111}$$

Thus, according to Lemma 3.13, we achieve not only global asymptotic stability in probability, but also inverse optimality.

Theorem 3.14 *The control law*

$$u^* = -\frac{2\beta}{3} M(x) z_n, \qquad \beta \geq 2 \qquad (3.112)$$

guarantees that the equilibrium at the origin of the system (3.28), (3.29) is globally asymptotically stable in probability and also minimizes the cost functional

$$J(u) = E\left\{\int_0^\infty \left[l(x) + \frac{27}{16\beta^2} M(x)^{-3} u^4\right] d\tau\right\}, \qquad (3.113)$$

for some positive definite radially unbounded function $l(x)$ parameterized by β.

We point out that the *quartic* form of the penalty on u in (3.113) is due to the *quartic* nature of the sclf.

3.6 Output-Feedback

3.6.1 Design for Output Feedback Systems

In this section we deal with nonlinear *output-feedback* systems driven by white noise. This class of systems is given by the following nonlinear stochastic differential equations:

$$\begin{aligned} dx_i &= x_{i+1} dt + \varphi_i(y)^T dw, \qquad i = 1, \cdots, n-1 \\ dx_n &= u\, dt + \varphi_n(y)^T dw \\ y &= x_1, \end{aligned} \qquad (3.114)$$

where $\varphi_i(y)$ are r-vector-valued smooth functions with $\varphi_i(0) = 0$, and w is an independent r-dimensional standard Wiener process.

Since the states $x_2, \cdots, x_n$ are not measured, we first design an observer which would provide exponentially convergent estimates of the unmeasured states in the absence of noise. The observer is designed as

$$\dot{\hat{x}}_i = \hat{x}_{i+1} + k_i(y - \hat{x}_1), \qquad i = 1, \cdots, n \qquad (3.115)$$

where $\hat{x}_{n+1} = u$. The observation errors $\tilde{x} = x - \hat{x}$ satisfy

$$\begin{aligned} d\tilde{x} &= \begin{bmatrix} -k_1 & & & \\ \vdots & & I & \\ -k_n & 0 & \cdots & 0 \end{bmatrix} \tilde{x}\, dt + \varphi(y)^T dw \\ &= A_0 \tilde{x}\, dt + \varphi(y)^T dw, \end{aligned} \qquad (3.116)$$

3.6 OUTPUT-FEEDBACK

where A_0 is designed to be asymptotically stable. Now, the entire system can be expressed as

$$\begin{aligned}
d\tilde{x} &= A_0 \tilde{x} dt + \varphi(y)^{\mathrm{T}} dw \\
dy &= (\hat{x}_2 + \tilde{x}_2) \, dt + \varphi_1(y)^{\mathrm{T}} dw \\
d\hat{x}_2 &= [\hat{x}_3 + k_2 (y - \hat{x}_1)] \, dt \\
&\vdots \\
d\hat{x}_n &= [u + k_n (y - \hat{x}_1)] \, dt.
\end{aligned} \tag{3.117}$$

Our output-feedback design will consist in applying a backstepping procedure to the system $(y, \hat{x}_2, \cdots, \hat{x}_n)$, while also taking care of the feedback connection through the $\tilde{x}$ system.

We employ a Lyapunov function of the form

$$V = \sum_{i=1}^{n} \frac{1}{4} z_i^4 + \left(\tilde{x}^{\mathrm{T}} P \tilde{x} \right)^2, \tag{3.118}$$

where P is a positive definite matrix which satisfies $A_0^{\mathrm{T}} P + P A_0 = -I$, and the error variables z_i are given by

$$\begin{aligned}
z_1 &= y & (3.119) \\
z_i &= \hat{x}_i - \alpha_{i-1} \left(\bar{\hat{x}}_{i-1}, y \right), \qquad i = 2, \cdots, n & (3.120)
\end{aligned}$$

using the yet-to-design stabilizing functions $\alpha_i(\bar{\hat{x}}_i, y)$, where $\bar{\hat{x}}_i = [\hat{x}_2, \cdots, \hat{x}_i]^{\mathrm{T}}$.

We start by an important preparatory comment. Since $\varphi_i(0) = 0$, the α_i's will vanish at $\bar{\hat{x}}_i = 0$, $y = 0$, as well as at $\bar{z}_i = 0$, where $\bar{z}_i = [z_1, \cdots, z_i]^{\mathrm{T}}$. Thus, by the mean value theorem, $\alpha_i(\bar{\hat{x}}_i, y)$ and $\varphi(y)$ can be expressed respectively as

$$\begin{aligned}
\alpha_i(\bar{\hat{x}}_i, y) &= \sum_{l=1}^{i} z_l \alpha_{il}(\bar{\hat{x}}_i, y), & (3.121) \\
\varphi(y) &= y \psi(y) & (3.122)
\end{aligned}$$

where $\alpha_{il}(\bar{\hat{x}}_i, y)$ and $\psi(y)$ are smooth functions.

Now, we are ready to start the backstepping design procedure. According to Itô's differentiation rule [86], we have

$$\begin{aligned}
dz_1 &= (\hat{x}_2 + \tilde{x}_2) \, dt + \varphi_1(y)^{\mathrm{T}} dw & (3.123) \\
dz_i &= \Bigg[\hat{x}_{i+1} + k_i \tilde{x}_1 - \sum_{l=2}^{i-1} \frac{\partial \alpha_{i-1}}{\partial \hat{x}_l} (\hat{x}_{l+1} + k_l \tilde{x}_1) - \frac{\partial \alpha_{i-1}}{\partial y} (\hat{x}_2 + \tilde{x}_2) \\
&\quad - \frac{1}{2} \left(\frac{\partial^2 \alpha_{i-1}}{\partial y^2} \right) \varphi_1(y)^{\mathrm{T}} \varphi_1(y) \Bigg] dt - \frac{\partial \alpha_{i-1}}{\partial y} \varphi_1(y)^{\mathrm{T}} dw & (3.124) \\
i &= 2, \cdots, n.
\end{aligned}$$

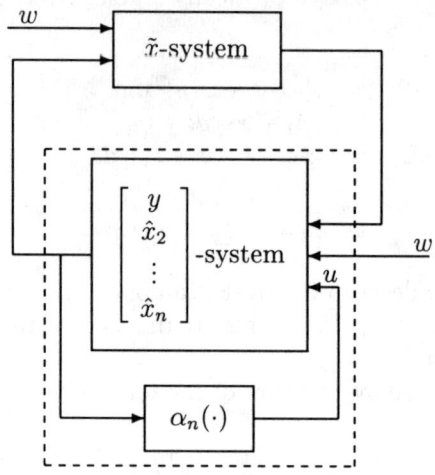

Figure 3.1: Feedback structure of the system (3.117)

As we announced previously, we employ a Lyapunov function of a quartic form

$$V(z,\tilde{x}) = \frac{1}{4}y^4 + \frac{1}{4}\sum_{i=2}^{n} z_i^4 + \frac{b}{2}\left(\tilde{x}^T P \tilde{x}\right)^2, \qquad (3.125)$$

where b is a positive constant. This form of the Lyapunov function clearly indicates that we view the system as a feedback connection in Figure 3.1. The first two terms in (3.125) constitute a Lyapunov function for the $(y, \hat{x}_2, \cdots, \hat{x}_n)$-system, while the third term in (3.125) is a Lyapunov function for the $\tilde{x}$-system. Even though not obvious from the calculations that follow, we achieve a nonlinear small-gain global stabilization (in probability) in the style of [54].

Now we start the process of selecting the functions $\alpha_i(\hat{\bar{x}}_i, y)$ to make $\mathcal{L}V$ negative definite. Along the solutions of (3.116), (3.123) and (3.124), we have

$$\begin{aligned}
\mathcal{L}V &= y^3\left(\hat{x}_2 + \tilde{x}_2\right) + \frac{3}{2}y^2 \varphi_1(y)^T \varphi_1(y) + \sum_{i=2}^{n} z_i^3 \left[\hat{x}_{i+1} + k_i \tilde{x}_1 - \sum_{l=2}^{i-1} \frac{\partial \alpha_{i-1}}{\partial \hat{x}_l}\right.\\
&\quad \left.\times (\hat{x}_{l+1} + k_l \tilde{x}_1) - \frac{\partial \alpha_{i-1}}{\partial y}(\hat{x}_2 + \tilde{x}_2) - \frac{1}{2}\left(\frac{\partial^2 \alpha_{i-1}}{\partial y^2}\right) \varphi_1(y)^T \varphi_1(y)\right]\\
&\quad + \frac{3}{2}\sum_{i=2}^{n} z_i^2 \left(\frac{\partial \alpha_{i-1}}{\partial y}\right)^2 \varphi_1(y)^T \varphi_1(y) - b\tilde{x}^T P \tilde{x} |\tilde{x}|^2\\
&\quad + 2b\,\mathrm{Tr}\left\{\varphi(y)\left(2P\tilde{x}\tilde{x}^T P + \tilde{x}^T P \tilde{x} P\right)\varphi(y)^T\right\}\\
&= -b\tilde{x}^T P \tilde{x} |\tilde{x}|^2 + 2b\,\mathrm{Tr}\left\{\varphi(y)\left(2P\tilde{x}\tilde{x}^T P + \tilde{x}^T P \tilde{x} P\right)\varphi(y)^T\right\}\\
&\quad + y^3(\alpha_1 + z_2 + \tilde{x}_2) + \frac{3}{2}y^2 \varphi_1(y)^T \varphi_1(y)
\end{aligned}$$

$$+ \sum_{i=2}^{n} z_i^3 \left[\alpha_i + z_{i+1} + k_i \tilde{x}_1 - \sum_{l=2}^{i-1} \frac{\partial \alpha_{i-1}}{\partial \hat{x}_l} (\hat{x}_{l+1} + k_l \tilde{x}_1) \right.$$

$$\left. - \frac{\partial \alpha_{i-1}}{\partial y} (\hat{x}_2 + \tilde{x}_2) - \frac{1}{2} \left(\frac{\partial^2 \alpha_{i-1}}{\partial y^2} \right) \varphi_1(y)^T \varphi_1(y) \right]$$

$$+ \frac{3}{2} \sum_{i=2}^{n} z_i^2 \left(\frac{\partial \alpha_{i-1}}{\partial y} \right)^2 \varphi_1(y)^T \varphi_1(y)$$

$$\leq - \left[b\lambda - 3bn\sqrt{n} \epsilon_2^2 |P|^4 - \frac{1}{4} \sum_{i=2}^{n} \frac{1}{\eta_i^4} - \frac{1}{4\epsilon_1^4} \right] |\tilde{x}|^4$$

$$+ y^3 \left[\alpha_1 + \frac{3}{2} \psi_1(y)^T \psi_1(y) y + \frac{3}{4} \delta_1^{\frac{4}{3}} y + \frac{3}{4} \epsilon_1^{\frac{4}{3}} y \right.$$

$$\left. + \frac{3}{4} \sum_{i=2}^{n} \xi_i^2 \left(\psi_1(y)^T \psi_1(y) \right)^2 y + \frac{3bn\sqrt{n}}{\epsilon_2^2} |\psi(y)|^4 y \right]$$

$$+ \sum_{i=2}^{n-1} z_i^3 \left[\alpha_i + k_i \tilde{x}_1 - \sum_{l=2}^{i-1} \frac{\partial \alpha_{i-1}}{\partial \hat{x}_l} (\hat{x}_{l+1} + k_l \tilde{x}_1) - \frac{\partial \alpha_{i-1}}{\partial y} \hat{x}_2 \right.$$

$$- \frac{1}{2} \frac{\partial^2 \alpha_{i-1}}{\partial y^2} \varphi_1(y)^T \varphi_1(y) + \frac{3}{4} \delta_i^{\frac{4}{3}} z_i + \frac{1}{4\delta_{i-1}^4} z_i$$

$$\left. + \frac{3}{4} \eta_i^{\frac{4}{3}} \left(\frac{\partial \alpha_{i-1}}{\partial y} \right)^{\frac{4}{3}} z_i + \frac{3}{4\xi_i^2} \left(\frac{\partial \alpha_{i-1}}{\partial y} \right)^4 z_i \right]$$

$$+ z_n^3 \left[u + k_n \tilde{x}_1 - \sum_{l=2}^{n-1} \frac{\partial \alpha_{n-1}}{\partial \hat{x}_l} (\hat{x}_{l+1} + k_l \tilde{x}_1) - \frac{\partial \alpha_{n-1}}{\partial y} \hat{x}_2 \right.$$

$$- \frac{1}{2} \frac{\partial^2 \alpha_{n-1}}{\partial y^2} \varphi_1(y)^T \varphi_1(y) + \frac{1}{4\delta_{n-1}^4} z_n$$

$$\left. + \frac{3}{4} \eta_n^{\frac{4}{3}} \left(\frac{\partial \alpha_{n-1}}{\partial y} \right)^{\frac{4}{3}} z_n + \frac{3}{4\xi_n^2} \left(\frac{\partial \alpha_{n-1}}{\partial y} \right)^4 z_n \right] \quad (3.126)$$

where $\lambda > 0$ is the smallest eigenvalue of P. The second equality comes from substituting $\hat{x}_i = z_i + \alpha_{i-1}$, and the inequality comes from the following Young's inequalities:

$$y^3 z_2 \leq \frac{3}{4} \delta_1^{\frac{4}{3}} y^4 + \frac{1}{4\delta_1^4} z_2^4 \quad (3.127)$$

$$y^3 \tilde{x}_2 \leq \frac{3}{4} \epsilon_1^{\frac{4}{3}} y^4 + \frac{1}{4\epsilon_1^4} \tilde{x}_2^4$$

$$\leq \frac{3}{4} \epsilon_1^{\frac{4}{3}} y^4 + \frac{1}{4\epsilon_1^4} |\tilde{x}|^4 \quad (3.128)$$

$$\sum_{i=2}^{n} z_i^3 z_{i+1} \leq \frac{3}{4} \sum_{i=2}^{n-1} \delta_i^{\frac{4}{3}} z_i^4 + \frac{1}{4} \sum_{i=3}^{n} \frac{1}{\delta_{i-1}^4} z_i^4 \quad (3.129)$$

$$-\sum_{i=2}^{n} z_i^3 \frac{\partial \alpha_{i-1}}{\partial y} \tilde{x}_2 \leq \frac{3}{4} \sum_{i=2}^{n} \eta_i^{\frac{4}{3}} \left(\frac{\partial \alpha_{i-1}}{\partial y}\right)^{\frac{4}{3}} z_i^4 + \frac{1}{4} \sum_{i=2}^{n} \frac{1}{\eta_i^4} \tilde{x}_2^4$$

$$\leq \frac{3}{4} \sum_{i=2}^{n} \eta_i^{\frac{4}{3}} \left(\frac{\partial \alpha_{i-1}}{\partial y}\right)^{\frac{4}{3}} z_i^4 + \frac{1}{4} \sum_{i=2}^{n} \frac{1}{\eta_i^4} |\tilde{x}|^4 \qquad (3.130)$$

$$\frac{3}{2} \sum_{i=2}^{n} z_i^2 \left(\frac{\partial \alpha_{i-1}}{\partial y}\right)^2 \varphi_1(y)^{\mathrm{T}} \varphi_1(y)$$

$$\leq \frac{3}{4} \sum_{i=2}^{n} \frac{1}{\xi_i^2} \left(\frac{\partial \alpha_{i-1}}{\partial y}\right)^4 z_i^4 + \frac{3}{4} \sum_{i=2}^{n} \xi_i^2 \left(\varphi_1(y)^{\mathrm{T}} \varphi_1(y)\right)^2 \qquad (3.131)$$

$$2b\mathrm{Tr}\left\{\varphi(y) \left(2P\tilde{x}\tilde{x}^{\mathrm{T}}P + \tilde{x}^{\mathrm{T}}P\tilde{x}P\right) \varphi(y)^{\mathrm{T}}\right\}$$

$$\leq 2bn \left|\varphi(y) \left(2P\tilde{x}\tilde{x}^{\mathrm{T}}P + \tilde{x}^{\mathrm{T}}P\tilde{x}P\right) \varphi(y)^{\mathrm{T}}\right|_{\infty}$$

$$\leq 2bn\sqrt{n} \left|\varphi(y) \left(2P\tilde{x}\tilde{x}^{\mathrm{T}}P + \tilde{x}^{\mathrm{T}}P\tilde{x}P\right) \varphi(y)^{\mathrm{T}}\right|$$

$$\leq 6bn\sqrt{n} y^2 |\psi(y)|^2 |P|^2 |\tilde{x}|^2 \qquad (\mathrm{cf.}\ (2.10))$$

$$\leq \frac{3bn\sqrt{n}}{\epsilon_2^2} y^4 |\psi(y)|^4 + 3bn\sqrt{n}\epsilon_2^2 |P|^4 |\tilde{x}|^4 \qquad (3.132)$$

At this point, we can see that all the terms can be cancelled by u and α_i. If we choose ϵ_1, ϵ_2 and η_i to satisfy

$$b\lambda - 3bn\sqrt{n}\epsilon_2^2 |P|^4 - \frac{1}{4} \sum_{i=2}^{n} \frac{1}{\eta_i^4} - \frac{1}{4\epsilon_1^4} = p > 0, \qquad (3.133)$$

and α_i and u as

$$\alpha_1 = -c_1 y - \frac{3}{2} \psi_1(y)^{\mathrm{T}} \psi_1(y) y - \frac{3}{4} \delta_1^{\frac{4}{3}} y - \frac{3}{4} \epsilon_1^{\frac{4}{3}} y - \frac{3}{4} \sum_{i=2}^{n} \xi_i^2 \left(\psi_1(y)^{\mathrm{T}} \psi_1(y)\right)^2 y$$

$$- \frac{3bn\sqrt{n}}{\epsilon_2^2} |\psi(y)|^4 y \qquad (3.134)$$

$$\alpha_i = -c_i z_i - k_i \tilde{x}_1 + \sum_{l=2}^{i-1} \frac{\partial \alpha_{i-1}}{\partial \hat{x}_l} (\hat{x}_{l+1} + k_l \tilde{x}_1) + \frac{\partial \alpha_{i-1}}{\partial y} \hat{x}_2$$

$$+ \frac{1}{2} \frac{\partial^2 \alpha_{i-1}}{\partial y^2} \varphi_1(y)^{\mathrm{T}} \varphi_1(y) - \frac{3}{4} \delta_i^{\frac{4}{3}} z_i - \frac{1}{4\delta_{i-1}^4} z_i$$

$$- \frac{3}{4} \eta_i^{\frac{4}{3}} \left(\frac{\partial \alpha_{i-1}}{\partial y}\right)^{\frac{4}{3}} z_i - \frac{3}{4\xi_i^2} \left(\frac{\partial \alpha_{i-1}}{\partial y}\right)^4 z_i \qquad (3.135)$$

$$u = -c_n z_n - k_n \tilde{x}_1 + \sum_{l=2}^{n-1} \frac{\partial \alpha_{n-1}}{\partial \hat{x}_l} (\hat{x}_{l+1} + k_l \tilde{x}_1) + \frac{\partial \alpha_{n-1}}{\partial y} \hat{x}_2$$

$$+ \frac{1}{2} \frac{\partial^2 \alpha_{n-1}}{\partial y^2} \varphi_1(y)^{\mathrm{T}} \varphi_1(y) - \frac{1}{4\delta_{n-1}^4} z_n$$

$$- \frac{3}{4} \eta_n^{\frac{4}{3}} \left(\frac{\partial \alpha_{n-1}}{\partial y}\right)^{\frac{4}{3}} z_n - \frac{3}{4\xi_n^2} \left(\frac{\partial \alpha_{n-1}}{\partial y}\right)^4 z_n, \qquad (3.136)$$

3.6 OUTPUT-FEEDBACK

where $c_i > 0$, then the infinitesimal generator of the closed-loop system (3.116), (3.123), (3.124) and (3.136) is negative definite:

$$\mathcal{L}V \leq -\sum_{i=1}^{n} c_i z_i^4 - p|\tilde{x}|^4. \tag{3.137}$$

With (3.137), we have the following stability result.

Theorem 3.15 *The equilibrium at the origin of the closed-loop stochastic system (3.117), (3.136) is globally asymptotically stable in probability.*

3.6.2 Inverse Optimal Output-Feedback Stabilization

From Lemma 3.13, we know that if we can design a stabilizing control law that has z_n as a factor, we can easily find another control law which solves the problem of inverse optimal stabilization in probability. If we consider carefully the last bracket of (3.126), every term except the second, the third, the fourth and the fifth has z_n as a factor. With the help of Young's inequalities

$$z_n^3 k_n \tilde{x}_1 \leq \frac{3}{4}\epsilon_3^{\frac{4}{3}} z_n^4 + \frac{1}{4\epsilon_3^4} k_n^4 \tilde{x}_1^4 \leq \frac{3}{4}\epsilon_3^{\frac{4}{3}} z_n^4 + \frac{1}{4\epsilon_3^4} k_n^4 |\tilde{x}|^4 \tag{3.138}$$

$$-z_n^3 \sum_{l=2}^{n-1} \frac{\partial \alpha_{n-1}}{\partial \hat{x}_l} k_l \tilde{x}_1 \leq \frac{3}{4}\left(\epsilon_4 \sum_{l=2}^{n-1} \frac{\partial \alpha_{n-1}}{\partial \hat{x}_l} k_l\right)^{\frac{4}{3}} z_n^4 + \frac{1}{4\epsilon_4^4} \tilde{x}_1^4$$

$$\leq \frac{3}{4}\left(\epsilon_4 \sum_{l=2}^{n-1} \frac{\partial \alpha_{n-1}}{\partial \hat{x}_l} k_l\right)^{\frac{4}{3}} z_n^4 + \frac{1}{4\epsilon_4^4} |\tilde{x}|^4 \tag{3.139}$$

$$-\frac{1}{2}z_n^3 \frac{\partial^2 \alpha_{n-1}}{\partial y^2} \varphi_1(y)^{\mathrm{T}} \varphi_1(y) = -\frac{1}{2}z_n^3 \frac{\partial^2 \alpha_{n-1}}{\partial y^2} \psi_1(y)^{\mathrm{T}} \psi_1(y) y^2$$

$$\leq \frac{3}{8}\left(\epsilon_5 \frac{\partial^2 \alpha_{n-1}}{\partial y^2} \psi_1(y)^{\mathrm{T}} \psi_1(y)\right)^{\frac{4}{3}} z_n^4 + \frac{1}{8\epsilon_5^4} y^4 \tag{3.140}$$

$$-z_n^3 \frac{\partial \alpha_{n-1}}{\partial y} \hat{x}_2 - z_n^3 \sum_{l=2}^{n-1} \frac{\partial \alpha_{n-1}}{\partial \hat{x}_l} \hat{x}_{l+1}$$

$$= -z_n^3 \sum_{l=2}^{n-1} \frac{\partial \alpha_{n-1}}{\partial \hat{x}_l} z_{l+1} - z_n^3 \frac{\partial \alpha_{n-1}}{\partial y} z_2 - z_n^3 \sum_{l=1}^{n-1} \sum_{k=1}^{l} \frac{\partial \alpha_{n-1}}{\partial \hat{x}_l} z_k \alpha_{lk}$$

(cf. (2.7), (2.9))

$$= -\sum_{l=2}^{n-1} z_n^3 \frac{\partial \alpha_{n-1}}{\partial \hat{x}_l} z_{l+1} - z_n^3 \frac{\partial \alpha_{n-1}}{\partial y} z_2 - \sum_{k=1}^{n-1} \sum_{l=k}^{n-1} \frac{\partial \alpha_{n-1}}{\partial \hat{x}_l} \alpha_{lk} z_n^3 z_k$$

$$\leq \sum_{l=2}^{n-1} \left[\frac{3}{4}\left(\epsilon_6 \frac{\partial \alpha_{n-1}}{\partial \hat{x}_l}\right)^{\frac{4}{3}} z_n^4 + \frac{1}{4\epsilon_6^4} z_{l+1}^4\right] + \frac{3}{4}\left(\epsilon_6 \frac{\partial \alpha_{n-1}}{\partial y}\right)^{\frac{4}{3}} z_n^4$$

$$+ \frac{1}{4\epsilon_6^4} z_2^4 + \sum_{k=1}^{n-1}\left[\frac{3}{4}\left(\epsilon_7 \sum_{l=k}^{n-1} \frac{\partial \alpha_{n-1}}{\partial \hat{x}_l} \alpha_{lk}\right)^{\frac{4}{3}} z_n^4 + \frac{1}{4\epsilon_7^4} z_k^4\right]$$

$$= z_n^4 \left[\frac{3}{4} \sum_{l=2}^{n-1} \left(\epsilon_6 \frac{\partial \alpha_{n-1}}{\partial \hat{x}_l} \right)^{\frac{4}{3}} + \frac{3}{4} \left(\epsilon_6 \frac{\partial \alpha_{n-1}}{\partial y} \right)^{\frac{4}{3}} + \frac{1}{4\epsilon_6^4} \right.$$
$$+ \frac{3}{4} \sum_{k=1}^{n-1} \left(\epsilon_7 \sum_{l=k}^{n-1} \frac{\partial \alpha_{n-1}}{\partial \hat{x}_l} \alpha_{lk} \right)^{\frac{4}{3}} \Bigg] + \sum_{i=2}^{n-1} \frac{1}{4\epsilon_6^4} z_i^4$$
$$+ \sum_{i=2}^{n-1} \frac{1}{4\epsilon_7^4} z_i^4 + \frac{1}{4\epsilon_7^4} y^4, \tag{3.141}$$

we have

$$\mathcal{L}V \leq -\left[b\lambda - 3bn\sqrt{n}\epsilon_2^2 |P|^4 - \frac{1}{4} \sum_{i=2}^{n} \frac{1}{\eta_i^4} - \frac{1}{4\epsilon_1^4} - \frac{1}{4\epsilon_4^4} - \frac{1}{4\epsilon_3^4} k_n^4 \right] |\tilde{x}|^4$$
$$+ y^3 \left[\alpha_1 + \frac{3}{2} \psi_1(y)^{\mathrm{T}} \psi_1(y) y + \frac{3}{4} \delta_1^{\frac{4}{3}} y \right.$$
$$\left. + \frac{3}{4} \epsilon_1^{\frac{4}{3}} y + \frac{3}{4} \sum_{i=2}^{n} \xi_i^2 \left(\psi_1(y)^{\mathrm{T}} \psi_1(y) \right)^2 y + \frac{3bn\sqrt{n}}{\epsilon_2^2} |\psi(y)|^4 y + \frac{1}{4\epsilon_7^4} y + \frac{1}{8\epsilon_5^4} y \right]$$
$$+ \sum_{i=2}^{n-1} z_i^3 \left[\alpha_i + k_i \tilde{x}_1 - \sum_{l=2}^{i-1} \frac{\partial \alpha_{i-1}}{\partial \hat{x}_l} (\hat{x}_{l+1} + k_l \tilde{x}_1) - \frac{\partial \alpha_{i-1}}{\partial y} \hat{x}_2 \right.$$
$$\left. - \frac{1}{2} \frac{\partial^2 \alpha_{i-1}}{\partial y^2} \varphi_1(y)^{\mathrm{T}} \varphi_1(y) + \frac{3}{4} \delta_i^{\frac{4}{3}} z_i + \frac{1}{4\delta_{i-1}^4} z_i \right.$$
$$\left. + \frac{3}{4} \eta_i^{\frac{4}{3}} \left(\frac{\partial \alpha_{i-1}}{\partial y} \right)^{\frac{4}{3}} z_i + \frac{3}{4\xi_i^2} \left(\frac{\partial \alpha_{i-1}}{\partial y} \right)^4 z_i + \frac{1}{4\epsilon_6^4} z_i + \frac{1}{4\epsilon_7^4} z_i \right]$$
$$+ z_n^3 \left[u + \frac{3}{4} \sum_{l=2}^{n-1} \left(\epsilon_6 \frac{\partial \alpha_{n-1}}{\partial \hat{x}_l} \right)^{\frac{4}{3}} z_n + \frac{3}{4} \left(\epsilon_6 \frac{\partial \alpha_{n-1}}{\partial y} \right)^{\frac{4}{3}} z_n + \frac{1}{4\epsilon_6^4} z_n \right.$$
$$\left. + \frac{3}{4} \sum_{k=1}^{n-1} \left(\epsilon_7 \sum_{l=k}^{n-1} \frac{\partial \alpha_{n-1}}{\partial \hat{x}_l} \alpha_{lk} \right)^{\frac{4}{3}} z_n + \frac{3}{8} \left(\epsilon_5 \frac{\partial^2 \alpha_{n-1}}{\partial y^2} \psi_1(y)^{\mathrm{T}} \psi_1(y) \right)^{\frac{4}{3}} z_n \right.$$
$$\left. + \frac{3}{4} \left(\epsilon_4 \sum_{l=2}^{n-1} \frac{\partial \alpha_{n-1}}{\partial \hat{x}_l} k_l \right)^{\frac{4}{3}} z_n + \frac{3}{4} \epsilon_3^{\frac{4}{3}} z_n + \frac{1}{4\delta_{n-1}^4} z_n + \frac{3}{4} \eta_n^{\frac{4}{3}} \left(\frac{\partial \alpha_{n-1}}{\partial y} \right)^{\frac{4}{3}} z_n \right.$$
$$\left. + \frac{3}{4\xi_n^2} \left(\frac{\partial \alpha_{n-1}}{\partial y} \right)^4 z_n \right]. \tag{3.142}$$

If $\epsilon_1, \epsilon_2, \epsilon_3, \epsilon_4, \epsilon_5, \epsilon_6, \epsilon_7$ and η_i are chosen to satisfy

$$b\lambda - 3bn\sqrt{n}\epsilon_2^2 |P|^4 - \frac{1}{4} \sum_{i=2}^{n} \frac{1}{\eta_i^4} - \frac{1}{4\epsilon_1^4} - \frac{1}{4\epsilon_4^4} - \frac{1}{4\epsilon_3^4} k_n^4 = p > 0 \tag{3.143}$$

$$\frac{1}{4\epsilon_7^4} + \frac{1}{8\epsilon_5^4} = \frac{c_1}{2} \tag{3.144}$$

$$\frac{1}{4\epsilon_6^4} + \frac{1}{4\epsilon_7^4} = \frac{c_i}{2}, \tag{3.145}$$

where c_1 and c_i are those in (3.134) and (3.135), and

$$u = -M(y, \hat{x}) z_n \tag{3.146}$$

$$\begin{aligned}M(y, \hat{x}) &= c_n + \frac{3}{4} \sum_{l=2}^{n-1} \left(\epsilon_6 \frac{\partial \alpha_{n-1}}{\partial \hat{x}_l} \right)^{\frac{4}{3}} + \frac{3}{4} \left(\epsilon_6 \frac{\partial \alpha_{n-1}}{\partial y} \right)^{\frac{4}{3}} + \frac{1}{4\epsilon_6^4} \\ &+ \frac{3}{4} \sum_{k=1}^{n-1} \left(\epsilon_7 \sum_{l=k}^{n-1} \frac{\partial \alpha_{n-1}}{\partial \hat{x}_l} \alpha_{lk} \right)^{\frac{4}{3}} + \frac{3}{8} \left(\epsilon_5 \frac{\partial^2 \alpha_{n-1}}{\partial y^2} \right)^{\frac{4}{3}} \\ &+ \frac{3}{4} \left(\epsilon_4 \sum_{l=2}^{n-1} \frac{\partial \alpha_{n-1}}{\partial \hat{x}_l} k_l \right)^{\frac{4}{3}} + \frac{3}{4} \epsilon_3^{\frac{4}{3}} + \frac{1}{4\delta_{n-1}^4} + \frac{3}{4} \eta_n^{\frac{4}{3}} \left(\frac{\partial \alpha_{n-1}}{\partial y} \right)^{\frac{4}{3}} \\ &+ \frac{3}{4 \xi_n^2} \left(\frac{\partial \alpha_{n-1}}{\partial y} \right)^4, \end{aligned} \tag{3.147}$$

with (3.134), (3.135) and (3.146), we get

$$\mathcal{L}V \le -\frac{1}{2} \sum_{i=1}^{n} c_i z_i^4 - p|\tilde{x}|^4. \tag{3.148}$$

Thus, according to Lemma 3.13, we achieve not only global asymptotic stability in probability, but also inverse optimality.

Theorem 3.16 *The control law*

$$u^* = -\beta M(y, \hat{x}) z_n, \qquad \beta \ge \frac{4}{3} \tag{3.149}$$

guarantees that the equilibrium at the origin of the system (3.114),(3.116) is globally asymptotically stable in probability and also minimizes the cost functional

$$J(u) = E\left\{ \int_0^\infty \left[l(x, \tilde{x}) + \frac{27}{16\beta^2} M(y, \hat{x})^{-3} u^4 \right] d\tau \right\}, \tag{3.150}$$

for some positive definite radially unbounded function $l(x, \tilde{x})$ parameterized by β.

3.6.3 Example

We give a second order example to illustrate Theorem 3.15. Consider the system

$$\begin{aligned} dx_1 &= x_2 dt + \frac{1}{2} x_1^2 dw \\ dx_2 &= u dt \\ y &= x_1. \end{aligned} \tag{3.151}$$

For this system, the estimator is

$$\begin{aligned}\dot{\hat{x}}_1 &= \hat{x}_2 + k_1(y - \hat{x}_1)\\ \dot{\hat{x}}_2 &= u + k_2(y - \hat{x}_1).\end{aligned} \qquad (3.152)$$

The virtual control α_1 and control u are:

$$\alpha_1 = -c_1 y - \frac{3}{8}y^3 - \frac{3}{4}\delta_1^{\frac{4}{3}}y - \frac{3}{4}\epsilon_1^{\frac{4}{3}}y - \frac{3}{64}\xi_2^2 y^5 - \frac{6\sqrt{2b}}{16\epsilon_2^2}y^5 \qquad (3.153)$$

$$\begin{aligned}u = &-c_2 z_2 - k_2\tilde{x}_1 + \frac{\partial \alpha_1}{\partial y}\hat{x}_2 + \frac{1}{8}\frac{\partial^2\alpha_1}{\partial y^2}y^4 - \frac{1}{4\delta_1^4}z_2\\ &-\frac{3}{4}\eta_2^{\frac{4}{3}}\left(\frac{\partial\alpha_1}{\partial y}\right)^{\frac{4}{3}}z_2 - \frac{3}{4\xi_2^2}\left(\frac{\partial\alpha_1}{\partial y}\right)^4 z_2\end{aligned} \qquad (3.154)$$

We choose $k_1 = 3$, $k_2 = 4.5$, $c_1 = 0.01$, $c_2 = 0.1$, $\delta_1 = 0.1$, $\xi_2 = 0.8$, $\epsilon_1 = 0.01$, $\eta_2 = 0.1$, $b = 0.1$, $\epsilon_2 = 50$, and set the initial condition at $x_1(0) = 1.3$, $x_2(0) = 0$, $\hat{x}_1(0) = 0$, $\hat{x}_2(0) = \alpha_1(0)$. The states and control of the system are shown in Figure 3.2. From Figure 3.2, we can see that the output converges to zero. It is also interesting to note how the solutions become less noisy as they approach zero—a consequence of the fact that the noise vector field vanishes at zero.

Notes and References

Even though an extensive coverage of stochastic Lyapunov theorems already exists in Khasminskii [59], Kushner [67], and Mao [74], in Section 3.1 the reader will find many refinements and improvements:

1. The first rigorous treatment of the *global* case (for example, compare the estimates in (3.15)–(3.20) with [59, 67, 74]);

2. A presentation based on class $\mathcal{K}$ functions rather than the more tedious ϵ-δ format [59, 67, 74] for the benefit of a student trained from Khalil [58];

3. A stochastic version of the convergence theorem due to LaSalle [68] and Yoshizawa [116].

Ever since the emergence of the stochastic stabilization theory in the 1960's [67], the progress has been plagued by a fundamental technical obstacle in the Lyapunov analysis—the Itô differentiation introduces not only the gradient but also the Hessian of the Lyapunov function. This diverted the attention from stabilization to optimization, including the risk-sensitive control problem [7, 25, 26, 50, 84, 95] and other problems [39, 40], effectively replacing

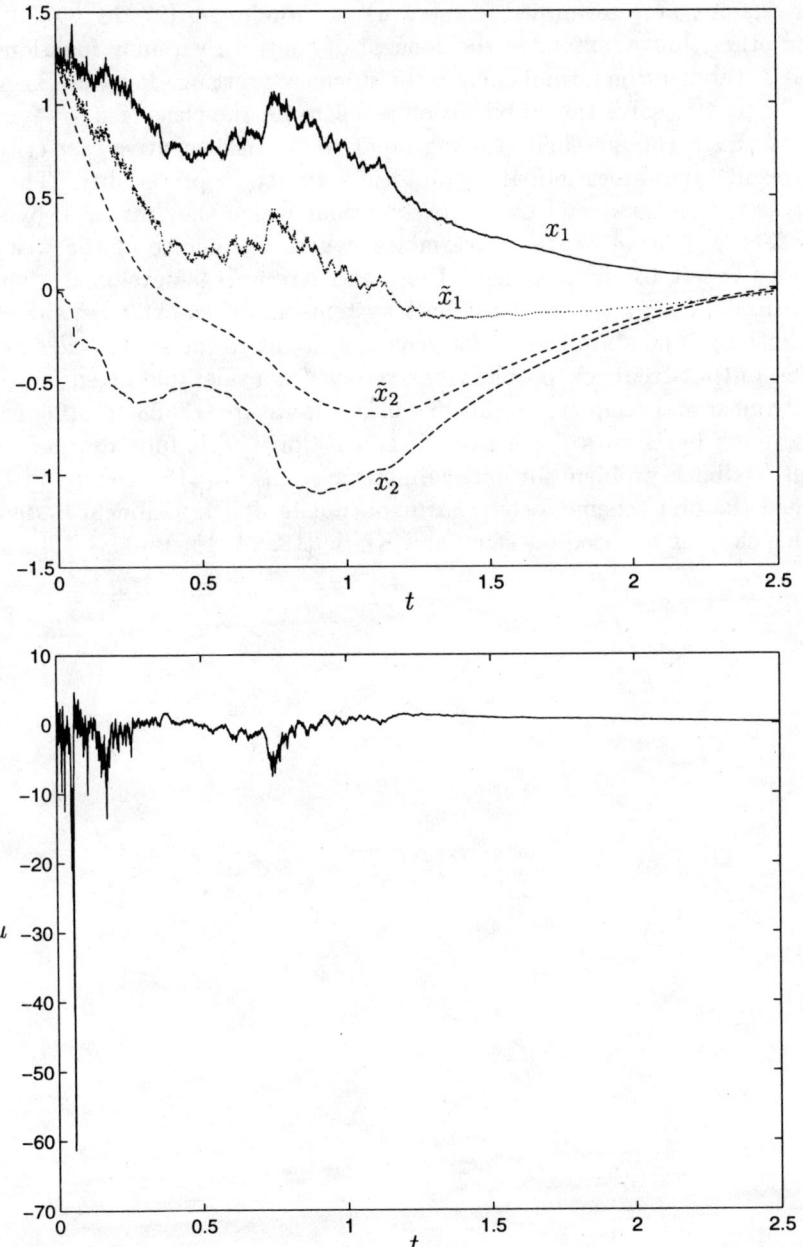

Figure 3.2: The states and control effort of the output-feedback system

the Lyapunov problem by an even more difficult problem of solving a Hamilton-Jacobi-Bellman PDE. Efforts toward (global) *stabilization* of stochastic nonlinear systems were revamped in the work of Florchinger [27, 28, 29, 30] who, among other things, extended the concept of control Lyapunov functions and Sontag's stabilization formula [99] to the stochastic setting. Pan and Basar [89] were the first to solve the stabilization problem for the class of strict-feedback systems. Even though their starting point was a risk-sensitive cost criterion, their result guarantees global asymptotic stability in probability. Their design is based on quadratic Lyapunov functions which they modified by state-dependent weighting on the z_i-variables, where the choice of the weighting functions is left to the designer. Deng and Krstic [14] developed a simpler (algorithmic) design for strict feedback systems and then extended the results on inverse optimal stabilization for general systems to the stochastic case [14].

The output-feedback problem has received considerable attention in the recent robust and adaptive nonlinear control literature [77, 63, 92, 106, 58, 52]. A new book by Byrnes, Delli Priscoli and Isidori [10] is fully devoted to the output feedback problem for deterministic systems. In [15], Deng and Krstić designed the first scheme for *stochastic* output feedback nonlinear systems.

This chapter is based on Deng and Krstic [13, 14, 15, 16].

Chapter 4

Stochastic Disturbance Attenuation

This chapter develops a stochastic counterpart of the disturbance attenuation results in Chapter 2. Unlike in Chapter 3,

1. the stochastic disturbance is allowed to drive the system in a way which prevents the existence of an equilibrium at the origin so (Lyapunov type) stability in probability is not achievable;

2. the designer is not given the knowledge of a bound on the intensity of the stochastic disturbance.

For example, linear systems driven (additively) by noise of unknown covariance fall into this category.

This set of assumptions leads to a stabilization problem statement in which we want to bound (in probability) the system solutions by the noise covariance, and to an optimal control problem given as a differential game played by control and the noise covariance as opposing players. Problems of this type have not been studied in previous literature on stochastic control.

4.1 Noise-to-State Stability (NSS)

In this section we extend the concept of input-to-state stability (ISS) from Section 2.1 to stochastic systems. ISS gives a relationship between $|x(t)|$ and $\sup_{0\leq\tau\leq t}|d(\tau)|$ in the system $\dot{x} = f(x) + g(x)d$. Clearly, for a stochastic system a characterization via $\sup_{0\leq\tau\leq t}|dw/d\tau|$ would not make sense because this quantity is not finite. For this reason, we propose a notion that relates $|x(t)|$ with a supremum of the incremental covariance of w, which is a statistical measure of the intensity of w. We refer to this notion as noise-to-state stability (NSS).

Consider the nonlinear stochastic system

$$dx = f(x)dt + g(x)dw, \qquad (4.1)$$

where $x \in \mathbb{R}^n$ is the state, w is an r-dimensional independent Wiener process with incremental covariance $\Sigma(t)\Sigma(t)^T dt$, i.e., $E\{dw dw^T\} = \Sigma(t)\Sigma(t)^T dt$, where $\Sigma(t)$ is a bounded function taking values in the set of nonnegative definite matrices, $f : \mathbb{R}^n \to \mathbb{R}^n$ and $g : \mathbb{R}^n \to \mathbb{R}^{n \times r}$ are locally Lipschitz, and $f(0) = 0$.

We first state notation which will be used in the sequel. For a matrix $X = [x_1, x_2, \cdots, x_n]$,

$$|X|_{\mathcal{F}} \triangleq \left(\text{Tr}\{X^T X\}\right)^{1/2} = \left(\text{Tr}\{X X^T\}\right)^{1/2} \qquad (4.2)$$

denotes the Frobenius norm, and obviously,

$$|X|_{\mathcal{F}} = |\text{col}(X)|, \qquad (4.3)$$

where $\text{col}(X) = [x_1^T, x_2^T, \cdots, x_n^T]^T$.

Definition 4.1 *The system (4.1) is noise-to-state stable (NSS) if $\forall \epsilon > 0$, there exists a class $\mathcal{KL}$ function $\beta(\cdot, \cdot)$ and a class $\mathcal{K}$ function $\gamma(\cdot)$, such that*

$$P\left\{|(x(t)| < \beta(|x_0|, t) + \gamma\left(\sup_{t \geq s \geq t_0} \left|\Sigma(s)\Sigma(s)^T\right|_{\mathcal{F}}\right)\right\} \geq 1 - \epsilon$$
$$\forall t \geq 0, \forall x_0 \in \mathbb{R}^n \setminus \{0\}. \qquad (4.4)$$

Theorem 4.2 *Consider system (4.1) and suppose there exists a C^2 function $V(x)$, class $\mathcal{K}_\infty$ functions α_1, α_2 and ρ, and a class $\mathcal{K}$ function α_3, such that*

$$\alpha_1(|x|) \leq V(x) \leq \alpha_2(|x|), \qquad (4.5)$$

and

$$|x| \geq \rho(|\Sigma\Sigma^T|_{\mathcal{F}})$$
$$\Downarrow \qquad (4.6)$$
$$\mathcal{L}V(x, \Sigma) = \frac{\partial V}{\partial x} f(x) + \frac{1}{2}\text{Tr}\left\{\Sigma^T g^T \frac{\partial^2 V}{\partial x^2} g \Sigma\right\} \leq -\alpha_3(|x|).$$

Then the system (4.1) is NSS.

Proof. Let $\tau \in [0, \infty)$ denote a time at which the trajectory enters the set $|x| \leq \rho\left(\sup_{s \geq t_0} |\Sigma(s)\Sigma(s)^T|_{\mathcal{F}}\right) = \rho(\|\Sigma\Sigma^T\|_\infty)$ for the first time. According to Theorem 3.3, for any $\epsilon' > 0$, there exists a class $\mathcal{KL}$ function β, such that

$$P\{|x(t)| < \beta(|x_0|, t)\} \geq 1 - \epsilon', \qquad \forall t \in [t_0, \tau), \forall x_0 \in \mathbb{R}^n \setminus \{0\}. \qquad (4.7)$$

Let us now turn our attention to the interval $t \in [\tau, \infty)$. Since

$$\frac{d}{dt}[EV(x(t))] = E\mathcal{L}V(x(t)), \tag{4.8}$$

which is negative for $x(t)$ outside the set $|x| \leq \rho\left(\|\Sigma\Sigma^T\|_\infty\right) \subseteq V(x) \leq \alpha_2\left(\rho\left(\|\Sigma\Sigma^T\|_\infty\right)\right)$, we have

$$EV(x(t)) \leq \alpha_2\left(\rho\left(\|\Sigma\Sigma^T\|_\infty\right)\right), \quad \forall t \geq \tau. \tag{4.9}$$

By Chebyshev's inequality, it follows that

$$P\left\{\sup_{t \in [\tau,\infty)} V(x(t)) \geq \delta\left(\alpha_2\left(\rho\left(\|\Sigma\Sigma^T\|_\infty\right)\right)\right)\right\}$$

$$\leq \frac{\alpha_2\left(\rho\left(\|\Sigma\Sigma^T\|_\infty\right)\right)}{\delta\left(\alpha_2\left(\rho\left(\|\Sigma\Sigma^T\|_\infty\right)\right)\right)} \leq \epsilon'', \tag{4.10}$$

where ϵ'' can be made arbitrarily small by an appropriate choice of $\delta \in \mathcal{K}_\infty$. Hence, for all $\epsilon'' > 0$, there exists $\gamma = \alpha_1^{-1} \circ \delta \circ \alpha_2 \circ \rho$ such that

$$P\left\{|x(t)| < \gamma\left(\|\Sigma\Sigma^T\|_\infty\right)\right\} \geq 1 - \epsilon'', \quad \forall t \geq \tau. \tag{4.11}$$

Thus, we get

$$P\left\{|x(t)| < \beta(|x_0|, t) + \gamma\left(\|\Sigma\Sigma^T\|_\infty\right)\right\}$$
$$\geq \max\{1 - \epsilon', 1 - \epsilon''\}$$
$$= 1 - \min\{\epsilon', \epsilon''\}$$
$$\triangleq 1 - \epsilon, \quad \forall t \geq t_0, \forall x_0 \in \mathbb{R}^n \setminus \{0\}. \tag{4.12}$$

By causality, it follows that

$$P\left\{|x(t)| < \beta(|x_0|, t) + \gamma\left(\sup_{t_0 \leq s \leq t}\left|\Sigma(s)\Sigma(s)^T\right|_\mathcal{F}\right)\right\} \geq 1 - \epsilon,$$
$$\forall t \geq t_0, \forall x_0 \in \mathbb{R}^n \setminus \{0\}. \tag{4.13}$$

$\square$

A function $V(x)$ that satisfies the conditions of Theorem 4.2 is referred to as the NSS-Lyapunov function.

4.2 Noise-to-State Stabilization and nss-clf's

Now we turn our attention to the system

$$dx = f(x)dt + g_1(x)dw + g_2(x)udt \tag{4.14}$$

where u is the control input, and study the problem of noise-to-state stabilizability by continuous feedback. We start with a definition which is a stochastic extension of ISS-control Lyapunov functions from Definition 2.3.

Definition 4.3 *A smooth positive definite radially unbounded function* $V : \mathbb{R}^n \to \mathbb{R}_+$ *is called an* **NSS-control Lyapunov function (nss-clf)** *for system (4.14), if there exists a class* $\mathcal{K}_\infty$ *function* ρ *such that the following implication holds for all* $x \neq 0$ *and* $\Sigma\Sigma^T \in \mathbb{R}^{r \times r}$:

$$|x| \geq \rho\left(\left|\Sigma\Sigma^T\right|_\mathcal{F}\right)$$
$$\Downarrow \qquad (4.15)$$
$$\inf_{u \in \mathbb{R}^m} \left\{ L_f V + \frac{1}{2}\mathrm{Tr}\left\{\Sigma^T g_1^T \frac{\partial^2 V}{\partial x^2} g_1 \Sigma\right\} + L_{g_2} V u \right\} < 0 \; .$$

Lemma 4.4 *A pair* (V, ρ), *where* $V(x)$ *is positive definite and radially unbounded and* $\rho \in \mathcal{K}_\infty$, *satisfies Definition 4.3 if and only if*

$$L_{g_2} V = 0 \;\Rightarrow\; L_f V(x) + \frac{1}{2}\left| g_1^T \frac{\partial^2 V}{\partial x^2} g_1 \right|_\mathcal{F} \rho^{-1}(|x|) < 0. \qquad (4.16)$$

Proof. (Necessity) By Definition 4.3, if $x \neq 0$ and $L_{g_2} V = 0$, then

$$|x| \geq \rho\left(\left|\Sigma\Sigma^T\right|_\mathcal{F}\right)$$
$$\Downarrow \qquad (4.17)$$
$$L_f V + \frac{1}{2}\mathrm{Tr}\left\{\Sigma^T g_1^T \frac{\partial^2 V}{\partial x^2} g_1 \Sigma\right\} < 0 \; .$$

Consider the incremental covariance given by the feedback law

$$\Sigma\Sigma^T = \rho^{-1}(|x|) \frac{g_1^T \frac{\partial^2 V}{\partial x^2} g_1}{\left|g_1^T \frac{\partial^2 V}{\partial x^2} g_1\right|_\mathcal{F}}, \qquad (4.18)$$

which satisfies the condition in (4.16):

$$\rho\left(\left|\Sigma\Sigma^T\right|_\mathcal{F}\right) = |x|. \qquad (4.19)$$

So, using (4.18),

$$\begin{aligned} L_f V + \frac{1}{2}\left| g_1^T \frac{\partial^2 V}{\partial x^2} g_1 \right|_\mathcal{F} \rho^{-1}(|x|) &= L_f V + \frac{1}{2}\mathrm{Tr}\left\{ g_1^T \frac{\partial^2 V}{\partial x^2} g_1 \Sigma\Sigma^T \right\} \\ &= L_f V + \frac{1}{2}\mathrm{Tr}\left\{\Sigma^T g_1^T \frac{\partial^2 V}{\partial x^2} g_1 \Sigma\right\} \\ &< 0. \qquad (4.20) \end{aligned}$$

(Sufficiency) For $|x| \geq \rho\left(\left|\Sigma\Sigma^T\right|_\mathcal{F}\right)$,

$$\inf_{u \in \mathbb{R}^m} \left\{ L_f V + \frac{1}{2}\mathrm{Tr}\left\{\Sigma^T g_1^T \frac{\partial^2 V}{\partial x^2} g_1 \Sigma\right\} + L_{g_2} V u \right\}$$

4.2 NOISE-TO-STATE STABILIZATION AND NSS-CLF'S

$$\leq \inf_{u \in \mathbb{R}^m} \left\{ L_f V + \frac{1}{2} \left| g_1^T \frac{\partial^2 V}{\partial x^2} g_1 \right|_{\mathcal{F}} \left| \Sigma \Sigma^T \right|_{\mathcal{F}} + L_{g_2} V u \right\}$$

$$\leq \inf_{u \in \mathbb{R}^m} \left\{ L_f V + \frac{1}{2} \left| g_1^T \frac{\partial^2 V}{\partial x^2} g_1 \right|_{\mathcal{F}} \rho^{-1}(|x|) + L_{g_2} V u \right\}$$

$$< 0 \tag{4.21}$$

□

We say that the system (4.14) is *noise-to-state stabilizable* if there exists a continuous feedback $u = \alpha(x)$ with $\alpha(0) = 0$ such that the system is noise-to-state stable.

Theorem 4.5 *The system (4.14) is noise-to-state stabilizable if there exists an nss-clf with small control property.*

Proof. Consider the Sontag type control law

$$\alpha_s(x) = \begin{cases} -\dfrac{\omega + \sqrt{\omega^2 + \left(L_{g_2} V (L_{g_2} V)^T\right)^2}}{L_{g_2} V (L_{g_2} V)^T} (L_{g_2} V)^T, & L_{g_2} V \neq 0 \\ 0, & L_{g_2} V = 0 \end{cases} \tag{4.22}$$

where

$$\omega = L_f V + \frac{1}{2} \left| g_1^T \frac{\partial^2 V}{\partial x^2} g_1 \right|_{\mathcal{F}} \rho^{-1}(|x|). \tag{4.23}$$

From Lemmas 4.4, 1.11 and 1.14, it follows that $\alpha_s(x)$ is continuous everywhere, so it remains to prove that it achieves NSS. Substituting (4.22) into $\mathcal{L}V$, we have

$$\mathcal{L}V = L_f V + \frac{1}{2} \text{Tr} \left\{ \Sigma^T g_1^T \frac{\partial^2 V}{\partial x^2} g_1 \Sigma \right\} + L_{g_2} V \alpha_s(x)$$

$$= -\sqrt{\omega^2 + \left(L_{g_2} V (L_{g_2} V)^T\right)^2} - \frac{1}{2} \left| g_1^T \frac{\partial^2 V}{\partial x^2} g_1 \right|_{\mathcal{F}} \left(\rho^{-1}(|x|) - \left| \Sigma \Sigma^T \right|_{\mathcal{F}} \right). \tag{4.24}$$

If $|x| \geq \rho\left(\left|\Sigma\Sigma^T\right|_{\mathcal{F}}\right)$, we have

$$\mathcal{L}V \leq -\sqrt{\omega^2 + \left(L_{g_2} V (L_{g_2} V)^T\right)^2} \triangleq -W(x) \tag{4.25}$$

where, according to Lemma 4.4, $W(x)$ is positive definite. By Theorem 4.2, the system is NSS.

□

4.3 Design for Strict-Feedback Systems

In this section we demonstrate how the concept of noise-to-state stability introduced in the last section can be achieved for a nontrivial class of stochastic nonlinear systems. We deal with strict-feedback systems driven by a stochastic process with time varying but bounded incremental covariance with an *unknown bound*. This class of systems is given by nonlinear stochastic differential equations

$$dx_i = x_{i+1}dt + \varphi_i(\bar{x}_i)^T dw, \quad i = 1, \cdots, n-1 \quad (4.26)$$
$$dx_n = udt + \varphi_n(\bar{x}_n)^T dw, \quad (4.27)$$

where $\varphi_i(\bar{x}_i)$ are r-vector-valued smooth functions, and w is an r-dimensional independent Wiener process with incremental covariance $\Sigma(t)\Sigma(t)^T dt$, i.e., $E\{dw dw^T\} = \Sigma(t)\Sigma(t)^T dt$, where $\Sigma(t)$ is a bounded function taking values in the set of nonnegative definite matrices.

To achieve noise-to-state stabilization, we employ the backstepping technique. We derive the stabilizing functions $\alpha_i(x_1, \cdots, x_i)$ simultaneously. We start with the transformation $z_i = x_i - \alpha_{i-1}$, and according to Itô's differentiation rule, we rewrite the system (4.26), (4.27) as

$$\begin{aligned} dz_i &= d(x_i - \alpha_{i-1}) \\ &= \left(x_{i+1} - \sum_{l=1}^{i-1} \frac{\partial \alpha_{i-1}}{\partial x_l} x_{l+1} - \frac{1}{2}\sum_{p,q=1}^{i-1} \frac{\partial^2 \alpha_{i-1}}{\partial x_p \partial x_q} \varphi_p^T \Sigma(t)\Sigma(t)^T \varphi_q\right) dt \\ &\quad + \left(\varphi_i^T - \sum_{l=1}^{i-1} \frac{\partial \alpha_{i-1}}{\partial x_l} \varphi_l^T\right) dw, \quad i = 1, \cdots, n, \end{aligned} \quad (4.28)$$

where $x_{n+1} = u$. We employ a Lyapunov function of the form

$$V(z) = \sum_{i=1}^n \frac{1}{4} z_i^4. \quad (4.29)$$

Now we set out to select the functions $\alpha_i(\bar{x}_i)$ to make $\mathcal{L}V \leq -cV + k|\Sigma|^4$, where c and k are positive constants. Along the solutions of (4.28), we have

$$\begin{aligned} \mathcal{L}V &= \sum_{i=1}^n z_i^3 \left(x_{i+1} - \sum_{l=1}^{i-1} \frac{\partial \alpha_{i-1}}{\partial x_l} x_{l+1} - \frac{1}{2}\sum_{p,q=1}^{i-1} \frac{\partial^2 \alpha_{i-1}}{\partial x_p \partial x_q} \varphi_p^T \Sigma\Sigma^T \varphi_q\right) \\ &\quad + \frac{3}{2}\sum_{i=1}^n z_i^2 \left(\varphi_i - \sum_{l=1}^{i-1} \frac{\partial \alpha_{i-1}}{\partial x_l}\varphi_l\right)^T \Sigma\Sigma^T \left(\varphi_i - \sum_{l=1}^{i-1} \frac{\partial \alpha_{i-1}}{\partial x_l}\varphi_l\right) \\ &\leq \sum_{i=1}^n z_i^3\left(\alpha_i - \sum_{l=1}^{i-1}\frac{\partial \alpha_{i-1}}{\partial x_l}x_{l+1}\right) + \sum_{i=1}^n z_i^3 z_{i+1} \end{aligned}$$

4.3 Design for Strict-Feedback Systems

$$+\frac{1}{2}\sum_{i=1}^{n}|z_i|^3\sum_{p,q=1}^{i-1}\left|\frac{\partial^2\alpha_{i-1}}{\partial x_p\partial x_q}\right||\varphi_p||\varphi_q|\left|\Sigma\Sigma^{\mathrm{T}}\right|$$

$$+\frac{3}{2}\sum_{i=1}^{n}z_i^2\left(\varphi_i-\sum_{l=1}^{i-1}\frac{\partial\alpha_{i-1}}{\partial x_l}\varphi_l\right)^{\mathrm{T}}\left(\varphi_i-\sum_{l=1}^{i-1}\frac{\partial\alpha_{i-1}}{\partial x_l}\varphi_l\right)\left|\Sigma\Sigma^{\mathrm{T}}\right|, \quad (4.30)$$

where $z_{n+1}=0$ and $\alpha_n=u$. Employing the inequality (4.42), we have

$$\begin{aligned}\mathcal{L}V \leq &\sum_{i=1}^{n}z_i^3\left(\alpha_i-\sum_{l=1}^{i-1}\frac{\partial\alpha_{i-1}}{\partial x_l}x_{l+1}\right)+\frac{3}{4}\sum_{i=1}^{n}\epsilon_i^{\frac{4}{3}}z_i^4+\sum_{i=1}^{n}\frac{1}{4\epsilon_{i-1}^4}z_i^4\\ &+\frac{1}{4}\sum_{i=1}^{n}z_i^6\sum_{p,q=1}^{i-1}\left(\frac{\partial^2\alpha_{i-1}}{\partial x_p\partial x_q}\right)^2\varphi_p^{\mathrm{T}}\varphi_p\varphi_q^{\mathrm{T}}\varphi_q+\frac{1}{4}\sum_{i=1}^{n}\sum_{p,q=1}^{i-1}\left|\Sigma\Sigma^{\mathrm{T}}\right|^2\\ &+\frac{3}{4}\sum_{i=1}^{n}z_i^4\left(\left(\varphi_i-\sum_{l=1}^{i-1}\frac{\partial\alpha_{i-1}}{\partial x_l}\varphi_l\right)^{\mathrm{T}}\left(\varphi_i-\sum_{l=1}^{i-1}\frac{\partial\alpha_{i-1}}{\partial x_l}\varphi_l\right)\right)+\frac{3}{4}\sum_{i=1}^{n}\left|\Sigma\Sigma^{\mathrm{T}}\right|^2\\ =&\sum_{i=1}^{n}z_i^3\left\{\alpha_i-\sum_{l=1}^{i-1}\frac{\partial\alpha_{i-1}}{\partial x_l}x_{l+1}+\frac{3}{4}\epsilon_i^{\frac{4}{3}}z_i+\frac{1}{4\epsilon_{i-1}^4}z_i\right.\\ &+\frac{1}{4}z_i^3\sum_{p,q=1}^{i-1}\left(\frac{\partial^2\alpha_{i-1}}{\partial x_p\partial x_q}\right)^2\varphi_p^{\mathrm{T}}\varphi_p\varphi_q^{\mathrm{T}}\varphi_q\\ &\left.+\frac{3}{4}z_i\left(\left(\varphi_i-\sum_{l=1}^{i-1}\frac{\partial\alpha_{i-1}}{\partial x_l}\varphi_l\right)^{\mathrm{T}}\left(\varphi_i-\sum_{l=1}^{i-1}\frac{\partial\alpha_{i-1}}{\partial x_l}\varphi_l\right)\right)^2\right\}\\ &+\left(\frac{(n-1)n(2n-1)}{24}+\frac{3}{4}n\right)|\Sigma|^4,\end{aligned}\quad(4.31)$$

where $\epsilon_n=0$ and $\epsilon_0=\infty$. Letting

$$\begin{aligned}\alpha_i =&-cz_i+\sum_{l=1}^{i-1}\frac{\partial\alpha_{i-1}}{\partial x_l}x_{l+1}-\frac{3}{4}\epsilon_i^{\frac{4}{3}}z_i-\frac{1}{4\epsilon_{i-1}^4}z_i\\ &-\frac{1}{4}z_i^3\sum_{p,q=1}^{i-1}\left(\frac{\partial^2\alpha_{i-1}}{\partial x_p\partial x_q}\right)^2\varphi_p^{\mathrm{T}}\varphi_p\varphi_q^{\mathrm{T}}\varphi_q\\ &-\frac{3}{4}z_i\left(\left(\varphi_i-\sum_{l=1}^{i-1}\frac{\partial\alpha_{i-1}}{\partial x_l}\varphi_l\right)^{\mathrm{T}}\left(\varphi_i-\sum_{l=1}^{i-1}\frac{\partial\alpha_{i-1}}{\partial x_l}\varphi_l\right)\right)^2\end{aligned}\quad(4.32)$$

$$u = \alpha_n \quad (4.33)$$

$$k = \frac{(n-1)n(2n-1)}{24}+\frac{3}{4}n, \quad (4.34)$$

we have

$$\mathcal{L}V \leq -cV+k\left|\Sigma\right|^4. \quad (4.35)$$

Since $V = \frac{1}{4}\sum_{i=1}^{n}[x_i - \alpha_{i-1}(\bar{x}_{i-1})]^4$ is positive definite and radially unbounded, there exists a class $\mathcal{K}_\infty$ function $\alpha(\cdot)$ such that $\mathcal{L}V \leq -\alpha(|x|) + k|\Sigma|^4$. With the choice $\rho(r) = \alpha^{-1}(2kr^4)$, the conditions of Theorem 4.2 are satisfied. Thus, the closed-loop system is NSS. Furthermore, from (4.35) we get $dEV/dt \leq -cEV + k|\Sigma|^4$, which implies $EV(t) \leq e^{-ct}EV(0) + k\int_0^t e^{-c(t-\tau)}|\Sigma(\tau)|^4 d\tau \leq e^{-ct}EV(0) + \frac{k}{c}\sup_{\tau \in [0,t]}|\Sigma(\tau)|^4$. The above conclusions are summarized in the following theorem.

Theorem 4.6 *The system (4.26), (4.27) with feedback (4.33) is NSS and, moreover,*

$$E\left\{|z(t)|_4^4\right\} \leq e^{-ct}|z(0)|_4^4 + \frac{4k}{c}\sup_{\tau \in [0,t]}|\Sigma(\tau)|^4. \tag{4.36}$$

where $|z|_4 = \left(\sum_i z_i^4\right)^{1/4}$.

The above development establishes that the Lyapunov function (4.29) is an nss-clf. This means that it can be employed in the formula (4.22) to derive a noise-to-state stabilizing control law alternative to (4.33). Since (4.33) is continuous, (4.29) satisfies the small control property, so (4.22) would result in a continuous control law. However, unlike (4.33), (4.22) would not be smooth.

When we set the nonlinearities $\varphi_i(\bar{x}_i)$ in (4.26), (4.27) to constant values, we get a linear system in the controllable canonical (chain of integrators) form. In this case the above procedure actually results in a linear control law. This is easy to see by noting that $\alpha_1(x_1)$ is linear, which inductively implies that the first partial derivatives of α_i are constant and that the second partial derivatives are zero. The linearity of the control law comes as somewhat of a surprise because of the quartic form of the Lyapunov function.

4.4 Inverse Optimal Noise-to-State Stabilization

In contrast to most of the work in stochastic nonlinear control where the starting point is an optimal (risk-sensitive) control problem [7, 25, 26, 50, 84, 88, 95], our approach in the previous sections was directed towards stability. In this section we establish connections with optimality. For general stochastic nonlinear systems (affine in control and noise) that are noise-to-state stabilizable, we design controllers that solve an inverse optimal control problem.

Consider the general nonlinear stochastic system affine in the noise w and control u:

$$dx = f(x)dt + g_1(x)dw + g_2(x)udt, \tag{4.37}$$

where w is an independent Wiener process with incremental covariance $\Sigma(t)\Sigma(t)^T dt$.

Definition 4.7 *The **inverse optimal stochastic gain assignment** problem for system (4.37) is solvable if there exist class $\mathcal{K}_\infty$ functions γ_1 and γ_2 whose derivatives γ_1' and γ_2' are also class $\mathcal{K}_\infty$ functions, a matrix-valued function $R_2(x)$ such that $R_2(x) = R_2(x)^T > 0$ for all x, a positive definite function $l(x)$, a positive definite radially unbounded function $S(x)$, and a continuous feedback control law $u = \alpha(x)$ with $\alpha(0) = 0$, which minimizes the cost functional*

$$J(u) = \sup_{\Sigma\Sigma^T \in \mathcal{D}} \left\{ \lim_{t\to\infty} E\left[S(x(t)) + \int_0^t \left(l(x) + \gamma_2\left(\left|R_2(x)^{1/2} u\right|\right) - \gamma_1\left(\left|\Sigma\Sigma^T\right|_{\mathcal{F}}\right) \right) d\tau \right] \right\}, \quad (4.38)$$

where $\mathcal{D}$ is the set of locally bounded functions of x.

This optimal control problem looks different than other problems considered in the literature. First, in the jargon of the risk-sensitive theory, (4.38) is a risk-neutral problem. Second, to see the main difference, consider the problem

$$I(u) = \lim_{t\to\infty} E\left[S(x(t)) + \int_0^t \left(l(x) + \gamma_2\left(\left|R_2(x)^{1/2} u\right|\right) \right) d\tau \right] \quad (4.39)$$

which appears as a direct nonlinear extension of the standard linear stochastic control problem [4] (a division by time t would lead to the optimal $\mathcal{H}_2$ problem [36]). This problem would be appropriate if Σ were constant and known. In that case the term $\int_0^t \gamma_1\left(\left|\Sigma\Sigma^T\right|_{\mathcal{F}}\right) d\tau$ would be included in the value function. However, when Σ is unknown and/or time varying, it is more reasonable to pose the problem as a differential game (4.38). (Further clarification is given in Remark 4.9). Note that this differential game is very different from stochastic differential games [7, Section 4.7.2] where the player opposed to control is another *deterministic* disturbance. In our case the opposing player is the stochastic disturbance w through its incremental covariance.

The next theorem allows a solution to the inverse optimal stochastic gain assignment problem provided a solution to a certain Hamilton-Jacobi-Isaacs equation is available. Before we state the theorem, we remind the reader of the Legendre-Fenchel transform. Let γ be a class $\mathcal{K}_\infty$ function whose derivative γ' is also a class $\mathcal{K}_\infty$ function; then $\ell\gamma$ denotes the L-F transform

$$\ell\gamma(r) = \int_0^r (\gamma')^{-1}(s) ds, \quad (4.40)$$

where $(\gamma')^{-1}(r)$ stands for the inverse function of $d\gamma(r)/dr$. The reader is referred to the Appendix A for some useful facts on the L-F transform.

Theorem 4.8 *Consider the control law*

$$u = \alpha(x) = -R_2^{-1} (L_{g_2}V)^{\mathrm{T}} \frac{\ell\gamma_2\left(\left|L_{g_2}VR_2^{-1/2}\right|\right)}{\left|L_{g_2}VR_2^{-1/2}\right|^2}, \qquad (4.41)$$

where $V(x)$ is a Lyapunov function candidate, γ_1 and γ_2 are class $\mathcal{K}_\infty$ functions whose derivatives are also class $\mathcal{K}_\infty$ functions, and $R_2(x)$ is a matrix-valued function such that $R_2(x) = R_2(x)^{\mathrm{T}} > 0$. If the control law (4.41) achieves global asymptotic stability in probability for the system

$$dx = f(x)dt + g_1(x)d\bar{w} + g_2(x)udt \qquad (4.42)$$

with respect to $V(x)$, where $\bar{w}$ is an independent r-dimensional stochastic process with incremental covariance

$$\bar{\Sigma}\bar{\Sigma}^{\mathrm{T}} = 2g_1^{\mathrm{T}} \frac{\partial^2 V}{\partial x^2} g_1 \frac{\ell\gamma_1\left(\left|g_1^{\mathrm{T}} \frac{\partial^2 V}{\partial x^2} g_1\right|_{\mathcal{F}}\right)}{\left|g_1^{\mathrm{T}} \frac{\partial^2 V}{\partial x^2} g_1\right|_{\mathcal{F}}^2}, \qquad (4.43)$$

then the control law

$$u^* = \alpha^*(x) = -\frac{\beta}{2} R_2^{-1} (L_{g_2}V)^{\mathrm{T}} \frac{(\gamma_2')^{-1}\left(\left|L_{g_2}VR_2^{-1/2}\right|\right)}{\left|L_{g_2}VR_2^{-1/2}\right|}, \quad \beta \geq 2 \qquad (4.44)$$

solves the problem of inverse optimal stochastic gain assignment for the system (4.37) by minimizing the cost functional

$$J(u) = \sup_{\Sigma \in \mathcal{D}} \left\{ \lim_{t \to \infty} E\left[2\beta V(x(t)) + \int_0^t \left(l(x) + \beta^2 \gamma_2\left(\frac{2}{\beta}\left|R_2^{1/2}u\right|\right)\right.\right.\right.$$
$$\left.\left.\left. -\beta\lambda\gamma_1\left(\frac{|\Sigma\Sigma^{\mathrm{T}}|_{\mathcal{F}}}{\lambda}\right)\right) d\tau \right] \right\}, \qquad (4.45)$$

where $\lambda \in (0, 2]$ and

$$l(x) = 2\beta\left[\ell\gamma_2\left(\left|L_{g_2}VR_2^{-1/2}\right|\right) - L_f V - \ell\gamma_1\left(\left|g_1^{\mathrm{T}}\frac{\partial^2 V}{\partial x^2}g_1\right|_{\mathcal{F}}\right)\right]$$
$$+\beta(\beta - 2)\ell\gamma_2\left(\left|L_{g_2}VR_2^{-1/2}\right|\right) + \beta(2-\lambda)\ell\gamma_1\left(\left|g_1^{\mathrm{T}}\frac{\partial^2 V}{\partial x^2}g_1\right|_{\mathcal{F}}\right). \qquad (4.46)$$

Remark 4.9 *Even though not explicit in the statement of Theorem 4.8, $V(x)$ solves the following family of Hamilton-Jacobi-Isaacs equations parameterized by $\beta \in [2, \infty)$ and $\lambda \in (0, 2]$:*

$$L_f V + \frac{\lambda}{2}\ell\gamma_1\left(\left|g_1^{\mathrm{T}}\frac{\partial^2 V}{\partial x^2}g_1\right|_{\mathcal{F}}\right) - \frac{\beta}{2}\ell\gamma_2\left(\left|L_{g_2}VR_2^{-1/2}\right|\right) + \frac{l(x)}{2\beta} = 0. \qquad (4.47)$$

4.4 INVERSE OPTIMAL NOISE-TO-STATE STABILIZATION

This equation, which depends only on known quantities, helps explain why we are pursuing a differential game problem with Σ as a player. If we set (4.39) as the goal, the resulting HJB equation is

$$L_f V + \frac{1}{2}\text{Tr}\left\{\Sigma^T g_1^T \frac{\partial^2 V}{\partial x^2} g_1 \Sigma\right\} - \frac{\beta}{2}\ell\gamma_2\left(\left|L_{g_2}V R_2^{-1/2}\right|\right) + \frac{l(x)}{2\beta} = 0. \quad (4.48)$$

If Σ is unknown (and allowed to take any value), it is clear that this equation cannot be solved. There is only one exception—linear systems. In the linear case $g_1(x)$ would be constant and $V(x)$ would be quadratic, which would make $g_1^T \frac{\partial^2 V}{\partial x^2} g_1$ constant. For a constant Σ, even if it is unknown, one would absorb the term $\frac{1}{2}\text{Tr}\left\{\Sigma^T g_1^T \frac{\partial^2 V}{\partial x^2} g_1 \Sigma\right\}$ into the value function. It is obvious that this can not be done when g_1 depends on x and/or $V(x)$ is nonquadratic. Thus, we pursue a differential game problem in which Σ is a player and its actions are penalized. $\square$

Proof of Theorem 4.8. Since the control law (4.41) globally asymptotically stabilizes the system (4.42), (4.43), there exists a continuous positive definite function $W : \mathbb{R}^n \to \mathbb{R}_+$ such that

$$\begin{aligned}
\mathcal{L}V\big|_{(4.41)} &= L_f V + \frac{1}{2}\text{Tr}\left\{\bar{\Sigma}^T g_1(x)^T \frac{\partial^2 V}{\partial x^2} g_1(x)\bar{\Sigma}\right\} + L_{g_2}V\alpha(x) \\
&= L_f V + \ell\gamma_1\left(\left|g_1^T \frac{\partial^2 V}{\partial x^2}g_1\right|_{\mathcal{F}}\right) - \ell\gamma_2\left(\left|L_{g_2}V R_2^{-1/2}\right|\right) \\
&\leq -W(x).
\end{aligned} \quad (4.49)$$

Then we have

$$\begin{aligned}
l(x) \geq\; & 2\beta W(x) + \beta(\beta-2)\ell\gamma_2\left(\left|L_{g_2}V R_2^{-1/2}\right|\right) \\
& + \beta(2-\lambda)\ell\gamma_1\left(\left|g_1^T\frac{\partial^2 V}{\partial x^2}g_1\right|_{\mathcal{F}}\right).
\end{aligned} \quad (4.50)$$

Since $W(x)$ is positive definite, $\beta \geq 2$, $\lambda \in (0, 2]$ and $\ell\gamma_2$ and $\ell\gamma_1$ are class $\mathcal{K}_\infty$ functions (Lemma A.1), $l(x)$ is positive definite. Therefore $J(u)$ is a meaningful cost functional.

Now we prove optimality. According to Dynkin's formula and by substituting $l(x)$ into $J(u)$, we have

$$\begin{aligned}
J(u) = \sup_{\Sigma \in \mathcal{D}} \Bigg\{ \lim_{t\to\infty} E\bigg[2\beta V(x(t)) &+ \int_0^t \bigg(l(x) + \beta^2\gamma_2\left(\frac{2}{\beta}\left|R_2^{1/2}u\right|\right) \\
&- \beta\lambda\gamma_1\left(\frac{\left|\Sigma\Sigma^T\right|_{\mathcal{F}}}{\lambda}\right)\bigg) d\tau\bigg]\Bigg\}
\end{aligned}$$

$$
\begin{aligned}
&= \sup_{\Sigma \in \mathcal{D}} \left\{ \lim_{t\to\infty} E\left[2\beta V(x(0)) + \int_0^t \left(2\beta \mathcal{L}V \,|_{(4.37)} + l(x) \right.\right.\right. \\
&\qquad\qquad \left.\left.\left. + \beta^2 \gamma_2 \left(\frac{2}{\beta} |R_2^{1/2} u|\right) - \beta \lambda \gamma_1 \left(\frac{|\Sigma\Sigma^{\mathrm{T}}|_{\mathcal{F}}}{\lambda}\right) \right) d\tau \right] \right\} \\
&= \sup_{\Sigma \in \mathcal{D}} \left\{ 2\beta E \{V(x(0))\} + \lim_{t\to\infty} E \int_0^t \left[\beta^2 \gamma_2 \left(\frac{2}{\beta} |R_2^{1/2} u|\right) \right.\right. \\
&\qquad\qquad + \beta^2 \ell_{\gamma_2} \left(|L_{g_2} V R_2^{-1/2}|\right) + 2\beta L_{g_2} V u - \beta \lambda \gamma_1 \left(\frac{|\Sigma\Sigma^{\mathrm{T}}|_{\mathcal{F}}}{\lambda}\right) \\
&\qquad\qquad \left.\left. - \beta \lambda \ell_{\gamma_1} \left(\left|g_1^{\mathrm{T}} \frac{\partial^2 V}{\partial x^2} g_1\right|_{\mathcal{F}}\right) + \beta \mathrm{Tr}\left\{\Sigma^{\mathrm{T}} g_1^{\mathrm{T}} \frac{\partial^2 V}{\partial x^2} g_1 \Sigma\right\} \right] d\tau \right\}. \quad (4.51)
\end{aligned}
$$

Using Lemma A.2 we have

$$
-2\beta L_{g_2} V u = \beta^2 \left(\frac{2}{\beta} R_2^{1/2} u\right)^{\mathrm{T}} \left(-R_2^{-1/2} (L_{g_2} V)^{\mathrm{T}}\right)
$$

$$
\leq \beta^2 \gamma_2 \left(\frac{2}{\beta} |R_2^{1/2} u|\right) + \beta^2 \ell_{\gamma_2} \left(|L_{g_2} V R_2^{-1/2}|\right) \quad (4.52)
$$

$$
\beta \mathrm{Tr}\left\{\Sigma^{\mathrm{T}} g_1^{\mathrm{T}} \frac{\partial^2 V}{\partial x^2} g_1 \Sigma\right\} = \beta \left(\mathrm{col}\left(\Sigma\Sigma^{\mathrm{T}}\right)\right)^{\mathrm{T}} \left(\mathrm{col}\left(g_1^{\mathrm{T}} \frac{\partial^2 V}{\partial x^2} g_1\right)\right)
$$

$$
\leq \beta \lambda \gamma_1 \left(\frac{|\Sigma\Sigma^{\mathrm{T}}|_{\mathcal{F}}}{\lambda}\right) + \beta \lambda \ell_{\gamma_1} \left(\left|g_1^{\mathrm{T}} \frac{\partial^2 V}{\partial x^2} g_1\right|_{\mathcal{F}}\right) \quad (4.53)
$$

and the equalities hold when

$$
u^* = -\frac{\beta}{2} R_2^{-1/2} (\gamma_2')^{-1} \left(|L_{g_2} V R_2^{-1/2}|\right) \frac{R_2^{-1/2} (L_{g_2} V)^{\mathrm{T}}}{|L_{g_2} V R_2^{-1/2}|} \quad (4.54)
$$

and

$$
\left(\Sigma\Sigma^{\mathrm{T}}\right)^* = \lambda (\gamma_1')^{-1} \left(\left|g_1^{\mathrm{T}} \frac{\partial^2 V}{\partial x^2} g_1\right|_{\mathcal{F}}\right) \frac{g_1^{\mathrm{T}} \frac{\partial^2 V}{\partial x^2} g_1}{\left|g_1^{\mathrm{T}} \frac{\partial^2 V}{\partial x^2} g_1\right|_{\mathcal{F}}}. \quad (4.55)
$$

So the "worst case" unknown covariance is given by (4.55), the minimum of (4.51) is reached with $u = u^*$, and

$$
\min_u J(u) = 2\beta E\{V(x(0))\}. \quad (4.56)
$$

To satisfy the requirements of Definition 4.7, it only remains to prove that $\alpha^*(x)$ is continuous and $\alpha^*(0) = 0$. This is proved in the proof of Theorem 3.9. □

The next theorem is the main result of this section. It constructs a controller that solves the problem posed in Definition 4.7.

4.4 INVERSE OPTIMAL NOISE-TO-STATE STABILIZATION

Theorem 4.10 *If the system (4.37) has an nss-clf $V(x)$ such that $g_1^{\mathrm{T}}\dfrac{\partial^2 V}{\partial x^2}g_1$ vanishes at the origin, then the problem of inverse optimal stochastic gain assignment is solvable.*

Proof. To solve the problem of inverse optimal stochastic gain assignment, we should find the functions $V(x)$, $R_2(x)$, $l(x)$, $\gamma_1(\cdot)$, $\gamma_2(\cdot)$ that solve the Hamilton-Jacobi-Isaacs equation (4.47) for some $\beta \in [2,\infty)$ and $\lambda \in (0,2]$. Then the inverse optimal controller would be given by (4.44). Since the system has an nss-clf, that is, there exists a pair (V,ρ) that satisfies Lemma 4.4, consider the choice

$$R_2(x) = I \begin{cases} \dfrac{2 L_{g_2} V \left(L_{g_2} V\right)^{\mathrm{T}}}{\omega + \sqrt{\omega^2 + \left(L_{g_2} V \left(L_{g_2} V\right)^{\mathrm{T}}\right)^2}}, & L_{g_2} V \neq 0 \\ \text{any positive number}, & L_{g_2} V = 0. \end{cases} \quad (4.57)$$

where ω is given by (4.23) and $\gamma_2(r) = \dfrac{1}{4}r^2$. In addition, let $\beta = \lambda = 2$, then $\ell\gamma_2(r) = r^2$, and after some computation we get

$$L_f V + \frac{\lambda}{2}\ell\gamma_1\left(\left|g_1^{\mathrm{T}}\frac{\partial^2 V}{\partial x^2}g_1\right|_{\mathcal{F}}\right) - \frac{\beta}{2}\ell\gamma_2\left(\left|L_{g_2} V R_2^{-1/2}\right|\right)$$
$$= -\frac{1}{2}\left[-\omega + \sqrt{\omega^2 + \left(L_{g_2} V \left(L_{g_2} V\right)^{\mathrm{T}}\right)^2}\right] - \frac{1}{2}\left|g_1^{\mathrm{T}}\frac{\partial^2 V}{\partial x^2}g_1\right|_{\mathcal{F}} \rho^{-1}(|x|)$$
$$+\ell\gamma_1\left(\left|g_1^{\mathrm{T}}\frac{\partial^2 V}{\partial x^2}g_1\right|_{\mathcal{F}}\right). \quad (4.58)$$

Since $g_1^{\mathrm{T}}\dfrac{\partial^2 V}{\partial x^2}g_1$ vanishes at the origin, there exists a class $\mathcal{K}_\infty$ function $\pi(|x|)$ such that $\left|g_1^{\mathrm{T}}\dfrac{\partial^2 V}{\partial x^2}g_1\right|_{\mathcal{F}} \leq \pi(|x|)$. Let $\zeta(r)$ be a class $\mathcal{K}_\infty$ function, whose derivative ζ' is also in $\mathcal{K}_\infty$, and such that $\zeta(r) \leq \dfrac{1}{2}r\rho^{-1}\left(\pi^{-1}(r)\right)$. Denoting $\gamma_1 = \ell\zeta$, since $\ell\ell\zeta = \zeta$, we have

$$\ell\gamma_1(r) = \zeta(r) \leq \frac{1}{2}r\rho^{-1}\left(\pi^{-1}(r)\right), \quad (4.59)$$

so

$$\ell\gamma_1\left(\left|g_1^{\mathrm{T}}\frac{\partial^2 V}{\partial x^2}g_1\right|_{\mathcal{F}}\right) \leq \frac{1}{2}\left|g_1^{\mathrm{T}}\frac{\partial^2 V}{\partial x^2}g_1\right|_{\mathcal{F}} \rho^{-1}(|x|). \quad (4.60)$$

Choose

$$l(x) = 4\left\{\frac{1}{2}\left[-\omega + \sqrt{\omega^2 + \left(L_{g_2} V \left(L_{g_2} V\right)^{\mathrm{T}}\right)^2}\right] + \frac{1}{2}\left|g_1^{\mathrm{T}}\frac{\partial^2 V}{\partial x^2}g_1\right|_{\mathcal{F}} \rho^{-1}(|x|)\right.$$

$$-\ell\gamma_1\left(\left|g_1^{\mathrm{T}}\frac{\partial^2 V}{\partial x^2}g_1\right|_{\mathcal{F}}\right)\Bigg\}$$
$$\geq 2\left[-\omega+\sqrt{\omega^2+\left(L_{g_2}V\left(L_{g_2}V\right)^{\mathrm{T}}\right)^2}\right], \tag{4.61}$$

which is positive definite by Lemma 4.4. Due to (4.58), this completes the selection of $V(x)$, $R_2(x)$, $l(x)$, $\gamma_1(\cdot)$, $\gamma_2(\cdot)$ that solve the HJI equation (4.47). □

Remark 4.11 The condition in Theorem 4.10 that $g_1^{\mathrm{T}}\frac{\partial^2 V}{\partial x^2}g_1$ be vanishing at the origin excludes the possibility of a linear system (g_1 =const) with a quadratic nss-clf $V(x)$. This condition can be eliminated by modifying the cost functional (4.38) but then other issues arise, like radial unboundedness of $\sqrt{\omega^2+(L_{g_2}V(L_{g_2}V)^{\mathrm{T}})^2}$. It is our opinion, supported by the results in Section 4.3, that, for stochastic systems, Lyapunov functions that are higher order at the origin are superior to quadratic Lyapunov functions. The peculiarity of the linear case (the fact that $\frac{1}{2}\mathrm{Tr}\left\{\Sigma^{\mathrm{T}}g_1^{\mathrm{T}}\frac{\partial^2 V}{\partial x^2}g_1\Sigma\right\}$ can be absorbed into the value function, making the controller independent of the noise vector field g_1!) has prevented the inadequacy of quadratic Lyapunov functions from being exposed for several decades now. □

Notes and References

The notion of noise-to-state stability is new but somewhat related to various ergodic concepts [19, 59]. Tsinias [108] recently introduced a different concept called "stochastic ISS" where he considers a system $dx = (f(x) + g_1(x)d)dt + g_2(x)dw$ (d is a deterministic input and w is a unity intensity Wiener process) and relates $|x(t)|$ to the deterministic quantity $\sup_{0\leq\tau\leq t}|d(t)|$ rather than to some measure of intensity of the stochastic input w.

As we explain throughout this chapter, the differential game/stochastic disturbance attenuation problem that we pursue in this chapter (in an inverse optimal setting) is different that any of the previous stochastic control problems—risk sensitive (or neutral), stochastic differential games, etc. The difference from the risk sensitive problem, in which Σ is fixed and known, is obvious. The difference from stochastic differential games is that, rather than keeping the covariance known/fixed and letting another *deterministic* disturbance be the opposing player, we leave the role of the opposing player to the covariance. This results in an energy-like bound $\int_0^\infty E\left[l(x)+\gamma_2\left(\left|R_2(x)^{1/2}u\right|\right)\right] \leq \int_0^\infty \gamma_1\left(\left|\Sigma\Sigma^{\mathrm{T}}\right|\right)$.

As we stated in Remarks 4.9 and 4.11, the approach in this chapter cures the anomaly in the LQG/$\mathcal{H}_2$ design where the controller does not depend on the noise input matrix B_1. A linear design that does take B_1 into account is *covariance control* [97], however, in covariance control, a bound on Σ needs to be known.

A major difficulty specific to the stochastic case is that $l(x)$ in Section 4.4 cannot be guaranteed to be radially unbounded as in the deterministic case in Theorem 2.10. Furthermore, unlike in Section 2.4, it is not clear how to make $R_2(x) = I$ which would allow deriving stability margins via passivity. The reason for both obstacles is the term $\partial^2 V/\partial x^2$ which prevents easy modifications of the Lyapunov function (in many cases this term acts to make $\mathcal{L}V$ less negative).

This chapter is based on Deng and Krstic [16].

Part III
ADAPTIVE CONTROL

Chapter 5

Deterministic Adaptive Tracking

Despite the fact that it deals with constant parametric uncertainties, the adaptive control problem is in many ways the most challenging of the stabilization problems treated in this book.

One of the main difficulties stems from the fact that the uncertain parameter vector is allowed to take an arbitrary value. For example, consider the scalar system $\dot{x} = u + x\theta$ where $\theta \in \mathbb{R}$ is unknown. The stabilization problem for this system cannot be solved via *static* feedback because the control law $u = \alpha(x)$ would have the impossible task to satisfy $x[\alpha(x) + x\theta] < 0$ for all $\theta \in \mathbb{R}$ and $x \neq 0$. Thus adaptive controllers require dynamic feedback. In its simplest form, this dynamic feedback involves a single scalar gain to be tuned as is the case in the next chapter. In a slightly more sophisticated form, the dynamic feedback involves an estimator of the uncertain parameter vector.

The constancy of the parametric uncertainty and the possibility to estimate it on line allow to pose more interesting and practically relevant stabilization problems than in the previous chapters. Instead of stabilization of an equilibrium, we can pursue the problem of trajectory tracking, i.e., the problem of stabilization of a feasible trajectory of the system.

5.1 The Adaptive Tracking Problem

Consider the system

$$\begin{aligned} \dot{x} &= f(x) + F(x)\theta + g(x)u \\ y &= h(x), \end{aligned} \tag{5.1}$$

where $x \in \mathbb{R}^n$, $u, y \in \mathbb{R}$, the mappings $f(x)$, $F(x)$, $g(x)$ and $h(x)$ are smooth, and θ is a constant *unknown* parameter vector which can take *any* value in $\mathbb{R}^p$. The quantity y is the output of the system which is required to track a pre-specified reference trajectory. In this chapter we study only the SISO problem (u, y scalar). The extension to the MIMO problem is straightforward.

To make tracking possible in the presence of an unknown parameter, we make the following key assumption.

Assumption 5.1 *For a given smooth function $y_r(t)$, there exist functions $\rho(t, \theta)$ and $\alpha_r(t, \theta)$ such that*

$$\begin{aligned} \frac{d\rho(t,\theta)}{dt} &= f(\rho(t,\theta)) + F(\rho(t,\theta))\theta + g(\rho(t,\theta))\alpha_r(t,\theta) \\ y_r(t) &= h(\rho(t,\theta)), \qquad \forall t \geq 0, \ \forall \theta \in \mathbb{R}^p. \end{aligned} \qquad (5.2)$$

Note that this implies that

$$\frac{\partial}{\partial \theta} h \circ \rho(t,\theta) = 0, \quad \forall t \geq 0, \ \forall \theta \in \mathbb{R}^p. \qquad (5.3)$$

For this reason, we can replace the objective of tracking the signal $y_r(t) = h \circ \rho(t, \theta)$ by the objective of tracking $y_r(t) = h \circ \rho(t, \hat{\theta}(t))$, where $\hat{\theta}(t)$ is a time function—an estimate of θ customary in adaptive control.

Consider the signal $x_r(t) = \rho(t, \hat{\theta}(t))$ which is governed by

$$\dot{x}_r = \frac{\partial \rho(t,\hat{\theta})}{\partial t} + \frac{\partial \rho(t,\hat{\theta})}{\partial \hat{\theta}} \dot{\hat{\theta}} = f(x_r) + F(x_r)\hat{\theta} + g(x_r)\alpha_r(t,\hat{\theta}) + \frac{\partial \rho(t,\hat{\theta})}{\partial \hat{\theta}} \dot{\hat{\theta}}. \qquad (5.4)$$

We define the tracking error $e = x - x_r = x - \rho(t, \hat{\theta})$ and compute its derivative:

$$\begin{aligned} \dot{e} &= f(x) - f(x_r) + [g(x) - g(x_r)]\alpha_r(t,\hat{\theta}) + F(x)\theta - F(x_r)\hat{\theta} \\ &\quad - \frac{\partial \rho(t,\hat{\theta})}{\partial \hat{\theta}}\dot{\hat{\theta}} + g(x)[u - \alpha_r(t,\hat{\theta})] \\ &= \tilde{f} + \tilde{F}\theta + F_r \tilde{\theta} - \frac{\partial \rho}{\partial \hat{\theta}}\dot{\hat{\theta}} + g\tilde{u}, \end{aligned} \qquad (5.5)$$

where $\tilde{\theta} = \theta - \hat{\theta}$ and

$$\begin{aligned} \tilde{f}(t,e,\hat{\theta}) &:= f(x) - f(x_r) + [g(x) - g(x_r)]\alpha_r(t,\hat{\theta}) \\ \tilde{F}(t,e,\hat{\theta}) &:= F(x) - F(x_r) \\ F_r(t,\hat{\theta}) &:= F(x_r) \\ \tilde{u} &:= u - \alpha_r(t,\hat{\theta}). \end{aligned} \qquad (5.6)$$

(With a slight abuse of notation, we will write $g(x)$ also as $g(t, e, \hat{\theta})$.) The global tracking problem is then transformed into the problem of global stabilization of the error system (5.5).

Definition 5.2 *The* **adaptive tracking** *problem for system (5.1) is solvable if Assumption 5.1 is satisfied and there exist a continuous function $\tilde{\alpha}(t, e, \hat{\theta})$ with $\tilde{\alpha}(t, 0, \hat{\theta}) \equiv 0$, a smooth function $\tau(t, e, \hat{\theta})$, and a positive definite symmetric $p \times p$ matrix Γ, such that the dynamic controller*

$$\tilde{u} = \tilde{\alpha}(t, e, \hat{\theta}) \tag{5.7}$$
$$\dot{\hat{\theta}} = \Gamma\tau(t, e, \hat{\theta}), \tag{5.8}$$

guarantees that the equilibrium $e = 0$, $\tilde{\theta} = 0$ of the system (5.5) is globally stable and $e(t) \to 0$ as $t \to \infty$ for any value of the unknown parameter $\theta \in \mathbb{R}^p$.

5.2 Stable Adaptive Tracking and atclf's

Our approach is to replace the problem of adaptive stabilization of (5.5) by a problem of non-adaptive stabilization of a modified system. This allows us to study adaptive stabilization in the framework of *control Lyapunov functions*.

Definition 5.3 *A smooth function $V_a : \mathbb{R}_+ \times \mathbb{R}^n \times \mathbb{R}^p \to \mathbb{R}_+$, positive definite, decrescent, and radially unbounded in e (uniformly in t) for each θ, is called an* **adaptive tracking control Lyapunov function (atclf)** *for (5.1) [or alternatively, an adaptive control Lyapunov function (aclf) for (5.5)], if Assumption 5.1 is satisfied and there exists a positive definite symmetric matrix $\Gamma \in \mathbb{R}^{p \times p}$ such that for each $\theta \in \mathbb{R}^p$, $V_a(t, e, \theta)$ is a* **clf** *for the modified non-adaptive system*

$$\dot{e} = \tilde{f} + \tilde{F}\theta + F\Gamma\left(\frac{\partial V_a}{\partial \theta}\right)^T - \frac{\partial \rho}{\partial \theta}\Gamma\left(\frac{\partial V_a}{\partial e}F\right)^T + g\tilde{u}, \tag{5.9}$$

that is, V_a satisfies

$$\inf_{\tilde{u} \in \mathbb{R}} \left\{ \frac{\partial V_a}{\partial t} + \frac{\partial V_a}{\partial e}\left[\tilde{f} + \tilde{F}\theta + F\Gamma\left(\frac{\partial V_a}{\partial \theta}\right)^T - \frac{\partial \rho}{\partial \theta}\Gamma\left(\frac{\partial V_a}{\partial e}F\right)^T + g\tilde{u}\right]\right\} < 0. \tag{5.10}$$

Lemma 5.4 *A function $V_a(t, e, \theta)$, positive definite, decrescent, and radially unbounded in $\mathcal{C}$ for each θ (uniformly in t), is an atclf for (5.1) if and only if, for all $e \neq 0$,*

$$\frac{\partial V_a}{\partial e}g = 0 \Longrightarrow \frac{\partial V_a}{\partial t} + \frac{\partial V_a}{\partial e}\left[\tilde{f} + \tilde{F}\theta + F\Gamma\left(\frac{\partial V_a}{\partial \theta}\right)^T - \frac{\partial \rho}{\partial \theta}\Gamma\left(\frac{\partial V_a}{\partial e}F\right)^T\right] < 0. \tag{5.11}$$

In the sequel we will show that in order to achieve adaptive stabilization of (5.5) it is necessary and sufficient to achieve non-adaptive stabilization of

(5.9). Noting that for $\tilde{\theta}(t) \equiv 0$ the system (5.5) reduces to the non-adaptive system

$$\dot{e} = \tilde{f} + \tilde{F}\theta + g\tilde{u}, \tag{5.12}$$

we see that the modification in (5.9) is

$$F\Gamma\left(\frac{\partial V_a}{\partial \theta}\right)^{\mathrm{T}} - \frac{\partial \rho}{\partial \theta}\Gamma\left(\frac{\partial V_a}{\partial e}F\right)^{\mathrm{T}}. \tag{5.13}$$

Since these terms are present only when Γ is nonzero, the role of these terms is to account for the effect of adaptation. Since $V_a(t, e, \theta)$ has a minimum at $e = 0$ for all t and θ, the modification terms vanish at $e = 0$, so $e = 0$ is an equilibrium of (5.9).

We now show how to design an adaptive controller (5.7)–(5.8) when an atclf is known.

Theorem 5.5 *The following two statements are equivalent:*

1. *There exists a triple $(\tilde{\alpha}, V_a, \Gamma)$ such that $\tilde{\alpha}(t, e, \theta)$ globally uniformly asymptotically stabilizes (5.9) at $e = 0$ for each $\theta \in \mathbb{R}^p$ with respect to the Lyapunov function $V_a(t, e, \theta)$.*

2. *There exists an atclf $V_a(t, e, \theta)$ for (5.1) with a small control property.*

Moreover, if an atclf $V_a(t, e, \theta)$ with a small control property exists, then the adaptive tracking problem for (5.1) is solvable.

Proof. (1 $\Rightarrow$ 2) Obvious because 1 implies that there exists a continuous function $W : \mathbb{R}_+ \times \mathbb{R}^n \times \mathbb{R}^p \to \mathbb{R}_+$, positive definite in e (uniformly in t) for each θ, such that

$$\frac{\partial V_a}{\partial t} + \frac{\partial V_a}{\partial e}\left[\tilde{f} + \tilde{F}\theta + F\Gamma\left(\frac{\partial V_a}{\partial \theta}\right)^{\mathrm{T}} - \frac{\partial \rho}{\partial \theta}\Gamma\left(\frac{\partial V_a}{\partial e}F\right)^{\mathrm{T}} + g\tilde{\alpha}\right] \leq -W(t, e, \theta). \tag{5.14}$$

Thus $V_a(t, e, \theta)$ is a clf for (5.9) for each $\theta \in \mathbb{R}^p$, and therefore it is an *atclf* for (5.1).

(2 $\Rightarrow$ 1) The proof of this part is based on Sontag's formula (1.28). This formula applied to (5.9) gives a control law:

$$\tilde{\alpha} = \begin{cases} -\dfrac{\dfrac{\partial V_a}{\partial t} + \dfrac{\partial V_a}{\partial e}\tilde{f} + \sqrt{\left(\dfrac{\partial V_a}{\partial t} + \dfrac{\partial V_a}{\partial e}\tilde{f}\right)^2 + \left(\dfrac{\partial V_a}{\partial e}g\right)^4}}{\dfrac{\partial V_a}{\partial e}g}, & \dfrac{\partial V_a}{\partial e}g \neq 0 \\ 0, & \dfrac{\partial V_a}{\partial e}g = 0 \end{cases} \tag{5.15}$$

5.2 STABLE ADAPTIVE TRACKING AND ATCLF'S

where

$$\bar{f} = \tilde{f} + \tilde{F}\theta + F\Gamma\left(\frac{\partial V_a}{\partial \hat\theta}\right)^{\mathrm{T}} - \frac{\partial \rho}{\partial \hat\theta}\Gamma\left(\frac{\partial V_a}{\partial e}F\right)^{\mathrm{T}}. \tag{5.16}$$

With the choice (5.15), inequality (5.14) is satisfied with the continuous function

$$W(t,e,\theta) = \sqrt{\left(\frac{\partial V_a}{\partial t}(t,e,\theta) + \frac{\partial V_a}{\partial e}\bar{f}(t,e,\theta)\right)^2 + \left(\frac{\partial V_a}{\partial e}g(t,e,\theta)\right)^4}, \tag{5.17}$$

which, by Lemma 5.4, is positive definite in e (uniformly in t) for each θ. By Lemmas 5.4 and 1.11, the control law (5.15) is smooth away from $e = 0$, and following the same argument as in Lemma 1.14, it is continuous at $e = 0$.

Assuming the existence of an atclf we now show that the adaptive tracking problem for (5.1) is solvable. Since $(2 \Rightarrow 1)$, there exists a triple $(\tilde{\alpha}, V_a, \Gamma)$ and a function W such that (5.14) is satisfied. Consider the Lyapunov function candidate

$$V(t,e,\hat\theta) = V_a(t,e,\hat\theta) + \frac{1}{2}(\theta - \hat\theta)^{\mathrm{T}}\Gamma^{-1}(\theta - \hat\theta). \tag{5.18}$$

With the help of (5.14), the derivative of V along the solutions of (5.5), (5.7), (5.8), is

$$\begin{aligned}
\dot{V} &= \frac{\partial V_a}{\partial t} + \frac{\partial V_a}{\partial e}\left[\tilde{f} + \tilde{F}\theta + F_r\tilde{\theta} - \frac{\partial \rho}{\partial \hat\theta}\Gamma\tau(t,e,\hat\theta) + g\tilde{\alpha}(t,e,\hat\theta)\right] \\
&\quad + \frac{\partial V_a}{\partial \hat\theta}\Gamma\tau(t,e,\hat\theta) - \tilde{\theta}^{\mathrm{T}}\tau(t,e,\hat\theta) \\
&= \frac{\partial V_a}{\partial t} + \frac{\partial V_a}{\partial e}\left[\tilde{f} + \tilde{F}\hat\theta + g\tilde{\alpha}(t,e,\hat\theta)\right] + \frac{\partial V_a}{\partial e}F\tilde{\theta} \\
&\quad - \frac{\partial V_a}{\partial e}\frac{\partial \rho}{\partial \hat\theta}\Gamma\tau + \frac{\partial V_a}{\partial \hat\theta}\Gamma\tau - \tilde{\theta}^{\mathrm{T}}\tau \\
&\leq -W(t,e,\hat\theta) - \frac{\partial V_a}{\partial \hat\theta}\Gamma\left(\frac{\partial V_a}{\partial e}F\right)^{\mathrm{T}} + \frac{\partial V_a}{\partial \hat\theta}\Gamma\tau + \frac{\partial V_a}{\partial e}\frac{\partial \rho}{\partial \hat\theta}\Gamma\left(\frac{\partial V_a}{\partial e}F\right)^{\mathrm{T}} \\
&\quad - \frac{\partial V_a}{\partial e}\frac{\partial \rho}{\partial \hat\theta}\Gamma\tau + \tilde{\theta}^{\mathrm{T}}\left(\frac{\partial V_a}{\partial e}F\right)^{\mathrm{T}} - \tilde{\theta}^{\mathrm{T}}\tau.
\end{aligned} \tag{5.19}$$

Choosing

$$\tau(t,e,\hat\theta) = \left(\frac{\partial V_a}{\partial e}F(t,e,\hat\theta)\right)^{\mathrm{T}}, \tag{5.20}$$

we get

$$\dot{V} \leq -W(t,e,\hat\theta). \tag{5.21}$$

Thus the equilibrium $e = 0, \tilde\theta = 0$ of (5.5), (5.7), (5.8) is globally stable, and by Theorem 1.5, $e(t) \to 0$ as $t \to \infty$. By Definition 5.2, the adaptive tracking problem for (5.1) is solvable. $\square$

The adaptive controller constructed in the proof of Theorem 5.5 consists of a control law $\tilde{u} = \tilde{\alpha}(t, e, \hat{\theta})$ given by (5.15), and an update law $\dot{\hat{\theta}} = \Gamma \tau(t, e, \hat{\theta})$ with (5.20). The control law $\tilde{\alpha}(t, e, \theta)$ is stabilizing for the modified system (5.9) but may not be stabilizing for the original system (5.5). However, as the proof of Theorem 5.5 shows, its certainty equivalence form $\tilde{\alpha}(t, e, \hat{\theta})$ is an adaptive globally stabilizing control law for the original system (5.5). The modified system "anticipates" parameter estimation transients, which results in incorporating the *tuning function* τ in the control law $\tilde{\alpha}$. Indeed, the formula (5.15) for $\tilde{\alpha}$ depends on τ via

$$\frac{\partial V_a}{\partial e}\bar{f}(t, e, \theta) = \frac{\partial V_a}{\partial e}(\bar{f} + \tilde{F}\theta) + \tau^{\mathrm{T}}\Gamma\left(\frac{\partial V_a}{\partial \theta} - \frac{\partial V_a}{\partial e}\frac{\partial \rho}{\partial \theta}\right)^{\mathrm{T}}, \qquad (5.22)$$

which is obtained by combining (5.16) and (5.20). Using (5.20) to rewrite the inequality (5.14) as

$$\frac{\partial V_a}{\partial t} + \frac{\partial V_a}{\partial e}\left[\bar{f} + \tilde{F}\theta + g\tilde{\alpha}(t, e, \theta)\right] + \left(\frac{\partial V_a}{\partial \theta} - \frac{\partial V_a}{\partial e}\frac{\partial \rho}{\partial \theta}\right)\Gamma\tau(t, e, \theta) \leq -W(t, e, \theta), \qquad (5.23)$$

it is not difficult to see that the control law (5.15) containing (5.22) prevents τ from destroying the nonpositiveness of the Lyapunov derivative.

Example 5.6 Consider the problem of designing an adaptive tracking controller for the system:

$$\begin{aligned}\dot{x}_1 &= x_2 + \varphi(x_1)^{\mathrm{T}}\theta \\ \dot{x}_2 &= u \\ y &= x_1.\end{aligned} \qquad (5.24)$$

In light of (5.1), $f(x) = [x_2, 0]^{\mathrm{T}}$, $F(x) = [\varphi(x_1), 0]^{\mathrm{T}}$, $g(x) = [0, 1]^{\mathrm{T}}$. For any given C^2 function $y_r(t)$, the function $\rho(t, \theta) = [\rho_1(t), \rho_2(t, \theta)]^{\mathrm{T}}$ is given by $\rho_1(t) = y_r(t)$ and $\rho_2(t, \theta) = \dot{y}_r(t) - \varphi(y_r)^{\mathrm{T}}\theta$, and the reference input is $\alpha_r(t, \theta) = \ddot{y}_r(t) - \frac{\partial \varphi(y_r)}{\partial y_r}^{\mathrm{T}}\theta \dot{y}_r(t)$. Hence Assumption 5.1 is satisfied.

With the signal $x_r(t) = \rho(t, \hat{\theta})$ and the tracking error $e = x - x_r$, we get the error system

$$\begin{aligned}\dot{e}_1 &= e_2 + \tilde{\varphi}^{\mathrm{T}}\theta + \varphi_r^{\mathrm{T}}\tilde{\theta} \\ \dot{e}_2 &= \tilde{u} - \frac{\partial \rho_2}{\partial \hat{\theta}}\dot{\hat{\theta}},\end{aligned} \qquad (5.25)$$

5.2 STABLE ADAPTIVE TRACKING AND ATCLF'S

where $\tilde{\varphi} = \varphi(x_1) - \varphi(x_{r1})$, $\varphi_r = \varphi(x_{r1}) = \varphi(y_r)$, $\tilde{u} = u - \alpha_r$. The modified non-adaptive error system is

$$\begin{aligned} \dot{e}_1 &= e_2 + \tilde{\varphi}^T \theta + \varphi^T \Gamma \left(\frac{\partial V_a}{\partial \theta} \right)^T \\ \dot{e}_2 &= \tilde{u} - \frac{\partial \rho_2}{\partial \theta} \Gamma \left(\frac{\partial V_a}{\partial e_1} \varphi^T \right)^T. \end{aligned} \qquad (5.26)$$

The control law $\tilde{u} = \tilde{\alpha}(t, e, \theta)$ with

$$\begin{aligned} \tilde{\alpha} = &-z_1 - c_2 z_2 - (-c_1 z_1 + z_2)\left(c_1 + \frac{\partial \tilde{\varphi}^T}{\partial e_1} \theta \right) - \frac{\partial \tilde{\varphi}^T}{\partial t}\theta \\ &- \varphi^T \Gamma \varphi \left[1, \; c_1 + \frac{\partial \tilde{\varphi}^T}{\partial e_1}\theta \right] z, \end{aligned} \qquad (5.27)$$

where $z_1 = e_1$, $z_2 = c_1 e_1 + e_2 + \tilde{\varphi}^T \theta$, and $c_1, c_2 > 0$, globally uniformly asymptotically stabilizes (5.26) at $e = 0$ with respect to $V_a = \frac{1}{2}(z_1^2 + z_2^2)$ with $W(t, e, \theta) = c_1 z_1^2 + c_2 z_2^2$. By Theorem 5.5, the adaptive tracking problem for (5.24) is solved with the control law $\tilde{u} = \tilde{\alpha}(t, e, \hat{\theta})$ and the update law

$$\dot{\hat{\theta}} = \Gamma \tau(t, e, \hat{\theta}) = \Gamma \varphi(t, e, \hat{\theta}) \left[1 \;\; c_1 + \frac{\partial \tilde{\varphi}^T}{\partial e_1} \hat{\theta} \right] z. \qquad (5.28)$$

$\square$

As it is always the case in adaptive control, in the proof of Theorem 5.5 we used a Lyapunov function $V(t, e, \hat{\theta})$ given by (5.18), which is quadratic in the parameter error $\theta - \hat{\theta}$. The quadratic form is suggested by the linear dependence of (5.5) on θ, and the fact that θ cannot be used for feedback. We will now show that the quadratic form of (5.18) is both necessary and sufficient for the existence of an atclf.

Definition 5.7 The **adaptive quadratic tracking** problem for (5.1) is solvable if the adaptive tracking problem for (5.1) is solvable and, in addition, there exist a smooth function $V_a(t, e, \theta)$ positive definite, decrescent, and proper in e (uniformly in t) for each θ, and a continuous function $W(t, e, \theta)$ positive definite in e (uniformly in t) for each θ, such that the derivative of (5.18) along the solutions of (5.5), (5.7), (5.8) is given by (5.21).

Corollary 5.8 The adaptive quadratic tracking problem for the system (5.1) is solvable if and only if there exists an atclf $V_a(t, e, \theta)$.

Proof. The 'if' part is contained in the proof of Theorem 5.5 where the Lyapunov function $V(t, e, \hat{\theta})$ is in the form (5.18). To prove the 'only if' part, we start by assuming global adaptive quadratic stabilizability of (5.5), and first show that $\tau(t,e,\hat{\theta})$ must be given by (5.20). The derivative of V along the solutions of (5.5), (5.7), (5.8), given by (5.19), is rewritten as

$$\begin{aligned}\dot{V} &= \frac{\partial V_a}{\partial t} + \frac{\partial V_a}{\partial e}\left[\tilde{f} + \tilde{F}\hat{\theta} + g\tilde{\alpha}(t,e,\hat{\theta})\right] - \frac{\partial V_a}{\partial e}\frac{\partial \rho}{\partial \hat{\theta}}\Gamma\tau(t,e,\hat{\theta}) + \frac{\partial V_a}{\partial \hat{\theta}}\Gamma\tau(t,e,\hat{\theta}) \\ &\quad - \hat{\theta}^{\mathrm{T}}\left(\left(\frac{\partial V_a}{\partial e}F\right)^{\mathrm{T}} - \tau\right) + \theta^{\mathrm{T}}\left(\left(\frac{\partial V_a}{\partial e}F\right)^{\mathrm{T}} - \tau\right).\end{aligned} \quad (5.29)$$

This expression has to be nonpositive to satisfy (5.21). Since it is affine in θ, it can be nonpositive for all $\theta \in \mathbb{R}^p$ only if the last term is zero, that is, only if τ is defined as in (5.20). Then, it is straightforward to verify that

$$\begin{aligned}&\frac{\partial V_a}{\partial t} + \frac{\partial V_a}{\partial e}\left[\tilde{f} + \tilde{F}\hat{\theta} + F\Gamma\left(\frac{\partial V_a}{\partial \hat{\theta}}\right)^{\mathrm{T}} - \frac{\partial \rho}{\partial \hat{\theta}}\Gamma\left(\frac{\partial V_a}{\partial e}F\right)^{\mathrm{T}} + g\tilde{\alpha}\right] \\ &= \dot{V} + \left(\tilde{\theta}^{\mathrm{T}} - \frac{\partial V_a}{\partial \hat{\theta}}\Gamma\right)\left(\tau - \left(\frac{\partial V_a}{\partial e}F\right)^{\mathrm{T}}\right) \\ &\leq -W(t,e,\hat{\theta})\end{aligned} \quad (5.30)$$

for all $(t,e,\hat{\theta}) \in \mathbb{R}_+ \times \mathbb{R}^{n+p}$. By $(1 \Rightarrow 2)$ in Theorem 5.5, $V_a(t,e,\theta)$ is an atclf for (5.1). $\square$

5.3 Adaptive Backstepping

With Theorem 5.5, the problem of adaptive stabilization is reduced to the problem of finding an atclf. This problem is solved recursively via backstepping.

Lemma 5.9 *If the adaptive quadratic tracking problem for the system*

$$\begin{aligned}\dot{x} &= f(x) + F(x)\theta + g(x)u \\ y &= h(x),\end{aligned} \quad (5.31)$$

is solvable with a C^1 control law, then the adaptive quadratic tracking problem for the augmented system

$$\begin{aligned}\dot{x} &= f(x) + F(x)\theta + g(x)\xi \\ \dot{\xi} &= u \\ y &= h(x),\end{aligned} \quad (5.32)$$

is also solvable.

5.3 ADAPTIVE BACKSTEPPING

Proof. Since the adaptive quadratic tracking problem for the system (5.31) is solvable, by Corollary 5.8 there exists an atclf $V_a(t, e, \theta)$ for (5.31), and by Theorem 5.5 it satisfies (5.14) with a control law $\tilde{\alpha}(t, e, \theta)$. Define $\tilde{\xi} = \xi - \alpha_r(t, \hat{\theta})$ and consider the system

$$\begin{aligned} \dot{e} &= \tilde{f}(t,e,\hat{\theta}) + \tilde{F}(t,e,\hat{\theta})\theta + F_r(t,\hat{\theta})\tilde{\theta} - \frac{\partial \rho}{\partial \hat{\theta}}\dot{\hat{\theta}} + g(t,e,\hat{\theta})\tilde{\xi} \\ \dot{\tilde{\xi}} &= \tilde{u} - \frac{\partial \alpha_r}{\partial \hat{\theta}}\dot{\hat{\theta}}, \end{aligned} \qquad (5.33)$$

where $\tilde{u} = u - \alpha_{r1}(t,\hat{\theta})$ and $\alpha_{r1}(t,\hat{\theta}) = \frac{\partial \alpha_r(t,\hat{\theta})}{\partial t}$. We will now show that

$$V_1(t,e,\tilde{\xi},\theta) = V_a(t,e,\theta) + \frac{1}{2}(\tilde{\xi} - \tilde{\alpha}(t,e,\theta))^2 \qquad (5.34)$$

is an atclf for the augmented system (5.32) by showing that it is a clf for the modified nonadaptive system

$$\begin{aligned} \dot{e} &= \tilde{f} + \tilde{F}\theta + F\Gamma\left(\frac{\partial V_1}{\partial \theta}\right)^T - \frac{\partial \rho}{\partial \theta}\Gamma\left(\frac{\partial V_1}{\partial e}F\right)^T + g\tilde{\xi} \\ \dot{\tilde{\xi}} &= \tilde{u} - \frac{\partial \alpha_r}{\partial \theta}\Gamma\left(\frac{\partial V_1}{\partial e}F\right)^T. \end{aligned} \qquad (5.35)$$

We present a constructive proof which shows that the control law

$$\begin{aligned} \tilde{u} &= \tilde{\alpha}_1(t,e,\tilde{\xi},\theta) \\ &= -c(\tilde{\xi} - \tilde{\alpha}) + \frac{\partial \tilde{\alpha}}{\partial t} - \frac{\partial V_a}{\partial e}g + \frac{\partial \tilde{\alpha}}{\partial e}\left(\tilde{f} + \tilde{F}\theta + g\tilde{\xi}\right) \\ &\quad + \left(\frac{\partial \alpha_r}{\partial \theta} + \frac{\partial \tilde{\alpha}}{\partial \theta} - \frac{\partial \tilde{\alpha}}{\partial e}\frac{\partial \rho}{\partial \theta}\right)\Gamma\left(\frac{\partial V_1}{\partial e}F\right)^T \\ &\quad + \left(\frac{\partial V_a}{\partial \theta} - \frac{\partial V_a}{\partial e}\frac{\partial \rho}{\partial \theta}\right)\Gamma\left(\frac{\partial \tilde{\alpha}}{\partial e}F\right)^T, \quad c > 0, \end{aligned} \qquad (5.36)$$

satisfies

$$\frac{\partial V_1}{\partial t} + \frac{\partial V_1}{\partial (e,\tilde{\xi})}\begin{bmatrix} \tilde{f} + \tilde{F}\theta + F\Gamma\left(\frac{\partial V_1}{\partial \theta}\right)^T - \frac{\partial \rho}{\partial \theta}\Gamma\left(\frac{\partial V_1}{\partial e}F\right)^T + g\tilde{\xi} \\ \tilde{\alpha}_1(t,e,\tilde{\xi},\theta) - \frac{\partial \alpha_r}{\partial \theta}\Gamma\left(\frac{\partial V_1}{\partial e}F\right)^T \end{bmatrix} \leq -W - c(\tilde{\xi}-\tilde{\alpha})^2. \qquad (5.37)$$

Let us start by introducing for brevity a new error state $z = \tilde{\xi} - \tilde{\alpha}(t,e,\theta)$. With (5.34) we compute

$$\begin{aligned} &\frac{\partial V_1}{\partial t} + \frac{\partial V_1}{\partial (e,\tilde{\xi})}\begin{bmatrix} \tilde{f} + \tilde{F}\theta + g\tilde{\xi} \\ \tilde{\alpha}_1(t,e,\tilde{\xi},\theta) \end{bmatrix} \\ &= \frac{\partial V_a}{\partial t} - z\frac{\partial \tilde{\alpha}}{\partial t} + \left(\frac{\partial V_a}{\partial e} - z\frac{\partial \tilde{\alpha}}{\partial e}\right)\left(\tilde{f} + \tilde{F}\theta + g\tilde{\xi}\right) + \frac{\partial V_1}{\partial \tilde{\xi}}\tilde{\alpha}_1 \end{aligned}$$

$$= \frac{\partial V_a}{\partial t} + \frac{\partial V_a}{\partial e}\left(\tilde{f} + \tilde{F}\theta + g\tilde{\alpha}\right)$$
$$+ z\left[\tilde{\alpha}_1 - \frac{\partial \tilde{\alpha}}{\partial t} + \frac{\partial V_a}{\partial e}g - \frac{\partial \tilde{\alpha}}{\partial e}\left(\tilde{f} + \tilde{F}\theta + g\tilde{\xi}\right)\right]. \tag{5.38}$$

On the other hand, in view of (5.34), we have

$$\frac{\partial V_1}{\partial (e,\tilde{\xi})}\left[\begin{array}{c} F\Gamma\left(\frac{\partial V_1}{\partial \theta}\right)^T - \frac{\partial \rho}{\partial \theta}\Gamma\left(\frac{\partial V_1}{\partial e}F\right)^T \\ -\frac{\partial \alpha_r}{\partial \theta}\Gamma\left(\frac{\partial V_1}{\partial e}F\right)^T \end{array}\right]$$
$$= \left(\frac{\partial V_a}{\partial e} - z\frac{\partial \tilde{\alpha}}{\partial e}\right)\left[F\Gamma\left(\frac{\partial V_a}{\partial \theta} - z\frac{\partial \tilde{\alpha}}{\partial \theta}\right)^T - \frac{\partial \rho}{\partial \theta}\Gamma\left(\frac{\partial V_a}{\partial e}F - z\frac{\partial \tilde{\alpha}}{\partial e}F\right)^T\right]$$
$$- z\frac{\partial \alpha_r}{\partial \theta}\Gamma\left(\frac{\partial V_1}{\partial e}F\right)^T$$
$$= \frac{\partial V_a}{\partial e}F\Gamma\left(\frac{\partial V_a}{\partial \theta}\right)^T - \frac{\partial V_a}{\partial e}\frac{\partial \rho}{\partial \theta}\Gamma\left(\frac{\partial V_a}{\partial e}F\right)^T$$
$$- z\left[\left(\frac{\partial \alpha_r}{\partial \theta} + \frac{\partial \tilde{\alpha}}{\partial \theta} - \frac{\partial \tilde{\alpha}}{\partial e}\frac{\partial \rho}{\partial \theta}\right)\Gamma\left(\frac{\partial V_1}{\partial e}F\right)^T \right.$$
$$\left. + \left(\frac{\partial V_a}{\partial \theta} - \frac{\partial V_a}{\partial e}\frac{\partial \rho}{\partial \theta}\right)\Gamma\left(\frac{\partial \tilde{\alpha}}{\partial e}F\right)^T\right]. \tag{5.39}$$

Adding (5.38) and (5.39), with (5.14) and (5.36) we get (5.37). This proves by Theorem 5.5 that $V_1(t,e,\tilde{\xi},\theta)$ is an atclf for system (5.32), or, an aclf for (5.33), and by Corollary 5.8 the adaptive quadratic tracking problem for this system is solvable. □

The new tuning function is determined by the new atclf V_1 and given by

$$\tau_1(t,e,\tilde{\xi},\theta) = \left(\frac{\partial V_1}{\partial (e,\tilde{\xi})}\begin{bmatrix} F \\ 0 \end{bmatrix}\right)^T = \left(\frac{\partial V_1}{\partial e}F\right)^T = \left[\left(\frac{\partial V_a}{\partial e} - (\tilde{\xi}-\tilde{\alpha})\frac{\partial \tilde{\alpha}}{\partial e}\right)F\right]^T$$
$$= \tau(t,e,\theta) - \left(\frac{\partial \tilde{\alpha}}{\partial e}F\right)^T(\tilde{\xi}-\tilde{\alpha}). \tag{5.40}$$

The control law $\tilde{\alpha}_1(t,e,\tilde{\xi},\theta)$ in (5.36) is only one out of many possible control laws. Once we have shown that V_1 given by (5.34) is an atclf for (5.32), or, an aclf for (5.33), we can use, for example, the C^0 control law $\tilde{\alpha}_1$ given by Sontag's formula (5.15).

Example 5.10 (Example 5.6, continued) Let us consider the system:

$$\begin{aligned}
\dot{x}_1 &= x_2 + \varphi(x_1)^{\mathrm{T}}\theta \\
\dot{x}_2 &= x_3 \\
\dot{x}_3 &= u \\
y &= x_1.
\end{aligned} \quad (5.41)$$

We treat the state x_3 as an integrator added to the (x_1, x_2)-subsystem for Example 5.6, so Lemma 5.9 is applicable. Defining $z_3 = \tilde{x}_3 - \tilde{\alpha}(t, e, \theta)$, where $\tilde{x}_3 = x_3 - \alpha_r$, by Lemma 5.9, the function $V_1(t, e, \tilde{x}_3, \theta) = \frac{1}{2}(z_1^2 + z_2^2 + z_3^2)$ is an atclf for the system (5.41). With (5.36) and (5.40) we obtain

$$\begin{aligned}
\tilde{u} &= \tilde{\alpha}_1(t, e, \tilde{\xi}, \theta) \\
&= -z_2 - cz_3 + \frac{\partial \tilde{\alpha}}{\partial t} + \frac{\partial \tilde{\alpha}}{\partial e}\begin{bmatrix} e_2 + \tilde{\varphi}^{\mathrm{T}}\theta \\ \tilde{x}_3 \end{bmatrix} \\
&\quad + \left(\frac{\partial \alpha_r}{\partial \theta} + \frac{\partial \tilde{\alpha}}{\partial \theta} + \frac{\partial \tilde{\alpha}}{\partial e_2}\varphi(y_r)^{\mathrm{T}}\right)\Gamma\tau_1 + \varphi^{\mathrm{T}}\Gamma\varphi\frac{\partial \tilde{\alpha}}{\partial e_1}z_2 \quad (5.42)
\end{aligned}$$

$$\tau_1 := \tau_1(t, e, \tilde{x}_3, \theta) = \tau - \frac{\partial \tilde{\alpha}}{\partial e_1}\varphi z_3. \quad (5.43)$$

The actual control is $u = \tilde{u} + \alpha_{r1}$ where $\alpha_{r1} = \dfrac{\partial \alpha_r}{\partial t} = \dddot{y}_r(t) - \dfrac{\partial \varphi(y_r)^{\mathrm{T}}}{\partial y_r}\theta \ddot{y}_r(t) - \dfrac{\partial^2 \varphi(y_r)^{\mathrm{T}}}{\partial y_r^2}\theta\left(\dot{y}_r(t)\right)^2$. □

A repeated application of Lemma 5.9 leads to the following result for a representative class of parametric strict-feedback systems.

Corollary 5.11 *The adaptive quadratic tracking problem for following system is solvable*

$$\begin{aligned}
\dot{x}_i &= x_{i+1} + \varphi_i(x_1, \ldots, x_i)^{\mathrm{T}}\theta, \quad i = 1, \ldots, n-1 \\
\dot{x}_n &= u + \varphi_n(x_1, \ldots, x_n)^{\mathrm{T}}\theta \\
y &= x_1.
\end{aligned} \quad (5.44)$$

5.4 Inverse Optimal Adaptive Tracking

While in the previous sections our objective was only to achieve adaptive tracking, in this section our objective is to achieve its optimality.

Definition 5.12 *The* **inverse optimal adaptive tracking** *problem for system (5.1) is solvable if there exist a positive constant β, a positive real-valued function $r(t, e, \theta)$, a real-valued function $l(t, e, \theta)$ positive definite in e for each θ, and a dynamic feedback law (5.7), (5.8) which solves the adaptive quadratic tracking problem and also minimizes the cost functional*

$$J = \beta \lim_{t \to \infty} |\theta - \hat{\theta}(t)|^2_{\Gamma^{-1}} + \int_0^\infty \left(l(t, e, \hat{\theta}) + r(t, e, \hat{\theta})\tilde{u}^2 \right) dt, \quad (5.45)$$

for any $\theta \in \mathbb{R}^p$.

This definition of optimality puts penalty on e and $\tilde{u}$ as well as on the terminal value of $|\tilde{\theta}|$. Even though $\tilde{\theta}(t)$ is not guaranteed to have a limit in the general tracking case (it is shown in Chapter 7 that $\tilde{\theta}(t)$ is guaranteed to have a limit in the case of set-point regulation), the existence of $\lim_{t \to \infty} |\tilde{\theta}|^2_{\Gamma^{-1}}$ is assured by the assumption that the adaptive *quadratic* tracking problem is solvable. This can be seen by noting that, since $V(t) \geq 0$ and from (5.21) $V(t)$ is nonincreasing, $V(t)$ has a limit. Since (5.21) guarantees that $V_a(t) \to 0$, it follows from (5.18) that $|\tilde{\theta}|^2_{\Gamma^{-1}}$ has a limit. The absence of an integral penalty on $\tilde{\theta}$ in (5.45) should not be surprising because adaptive feedback systems, in general, do not guarantee parameter convergence to a true value.

Theorem 5.13 *Suppose there exists an atclf $V_a(t, e, \theta)$ for (5.1) and a control law $\tilde{u} = \tilde{\alpha}(t, e, \theta)$ that stabilizes the system*

$$\dot{e} = \underbrace{\tilde{f} + \tilde{F}\theta + F\Gamma\left(\frac{\partial V_a}{\partial \theta}\right)^{\mathrm{T}} - \frac{\partial \rho}{\partial \theta}\Gamma\left(\frac{\partial V_a}{\partial e}F\right)^{\mathrm{T}}}_{\bar{f}} + g\tilde{u} \quad (5.46)$$

has the form

$$\tilde{\alpha}(t, e, \theta) = -r^{-1}(t, e, \theta)\frac{\partial V_a}{\partial e}g, \quad (5.47)$$

where $r(t, e, \theta) > 0$ for all t, e, θ. Then

1. *The non-adaptive control law*

$$\tilde{u} = \tilde{\alpha}^*(t, e, \theta) = \beta\tilde{\alpha}(t, e, \theta), \quad \beta \geq 2, \quad (5.48)$$

minimizes the cost functional

$$J_a = \int_0^\infty \left(l(t, e, \theta) + r(t, e, \theta)\tilde{u}^2 \right) dt, \quad \forall \theta \in \mathbb{R}^p \quad (5.49)$$

along the solutions of the nonadaptive system (5.46), where

$$l(t, e, \theta) = -2\beta \left[\frac{\partial V_a}{\partial t} + \frac{\partial V_a}{\partial e}(\bar{f} + g\tilde{\alpha}) \right] + \beta(\beta - 2)r^{-1}\left(\frac{\partial V_a}{\partial e}g\right)^2. \quad (5.50)$$

5.4 INVERSE OPTIMAL ADAPTIVE TRACKING

2. *The inverse optimal adaptive tracking problem is solvable.*

Proof. (Part 1) In light of (5.14), we have

$$\begin{aligned}
l(t,e,\theta) &= -2\beta\left[\frac{\partial V_a}{\partial t} + \frac{\partial V_a}{\partial e}(\bar{f}+g\tilde{\alpha})\right] + \beta(\beta-2)r^{-1}\left(\frac{\partial V_a}{\partial e}g\right)^2 \\
&\geq 2\beta W(t,e,\theta) + \beta(\beta-2)r^{-1}\left(\frac{\partial V_a}{\partial e}g\right)^2.
\end{aligned} \quad (5.51)$$

Since $\beta \geq 2$, $r(t,e,\theta) > 0$, and $W(t,e,\theta)$ is positive definite, $l(t,e,\theta)$ is also positive definite. Therefore J_a defined in (5.49) is a meaningful cost functional, which puts penalty both on e and $\tilde{u}$. Substituting $l(t,e,\theta)$ and

$$v = \tilde{u} - \tilde{\alpha}^* = \tilde{u} + \beta r^{-1}\frac{\partial V_a}{\partial e}g \quad (5.52)$$

into J_a, we get

$$\begin{aligned}
J_a &= \int_0^\infty \left[-2\beta\frac{\partial V_a}{\partial t} - 2\beta\frac{\partial V_a}{\partial e}\bar{f} + \beta^2 r^{-1}\left(\frac{\partial V_a}{\partial e}g\right)^2 \right. \\
&\qquad \left. + rv^2 - 2\beta v\frac{\partial V_a}{\partial e}g + \beta^2 r^{-1}\left(\frac{\partial V_a}{\partial e}g\right)^2\right] dt \\
&= -2\beta \int_0^\infty \left[\frac{\partial V_a}{\partial t} + \frac{\partial V_a}{\partial e}(\bar{f}+g\tilde{u})\right] dt + \int_0^\infty rv^2 dt \\
&= -2\beta \int_0^\infty dV_a + \int_0^\infty rv^2 dt \\
&= 2\beta V_a(0,e(0),\hat{\theta}(0)) - 2\beta \lim_{t\to\infty} V_a(t,e(t),\hat{\theta}(t)) + \int_0^\infty rv^2 dt. \quad (5.53)
\end{aligned}$$

Since the control input $\tilde{u}(t)$ solves the adaptive quadratic tracking problem, $\lim_{t\to\infty} e(t) = 0$, and we have that $\lim_{t\to\infty} V_a(t,e(t),\hat{\theta}(t)) = 0$. Therefore, the minimum of (5.53) is reached only if $v = 0$, and hence the control $\tilde{u} = \tilde{\alpha}^*(t,e,\theta)$ is an optimal control.

(Part 2) Since there exists an atclf V_a for (5.1), the adaptive quadratic tracking problem is solvable. Next, we show that the dynamic control law

$$\tilde{u} = \tilde{\alpha}^*(t,e,\hat{\theta}) \quad (5.54)$$

$$\dot{\hat{\theta}} = \Gamma\tau(t,e,\hat{\theta}) = \Gamma\left(\frac{\partial V_a}{\partial e}F\right)^T \quad (5.55)$$

minimizes the cost functional (5.45). The choice of the update law (5.55) is due to the requirement that (5.54), (5.55) solves the adaptive *quadratic* tracking problem (see the proof of Corollary 5.8). Substituting $l(t,e,\hat{\theta})$ and

$$v = \tilde{u} + \beta r^{-1}\frac{\partial V_a}{\partial e}g \quad (5.56)$$

into J, along the solutions of (5.5) and (5.55) we get

$$\begin{aligned}
J &= \beta \lim_{t\to\infty} |\tilde{\theta}|^2_{\Gamma^{-1}} + \int_0^\infty \left[-2\beta \frac{\partial V_a}{\partial t} \right.\\
&\quad -2\beta \frac{\partial V_a}{\partial e}\left(\tilde{f} + \tilde{F}\theta + F_r\tilde{\theta} - \frac{\partial \rho}{\partial \hat{\theta}}\Gamma\left(\frac{\partial V_a}{\partial e}F\right)^{\mathrm{T}} \right) \\
&\quad - 2\beta \left(\frac{\partial V_a}{\partial \hat{\theta}} - \tilde{\theta}^{\mathrm{T}}\Gamma^{-1} \right)\Gamma\left(\frac{\partial V_a}{\partial e}F\right)^{\mathrm{T}} + \beta^2 r^{-1}\left(\frac{\partial V_a}{\partial e}g\right)^2 \\
&\quad \left. +rv^2 - 2\beta v \frac{\partial V_a}{\partial e}g + \beta^2 r^{-1}\left(\frac{\partial V_a}{\partial e}g\right)^2 \right] dt \\
&= \beta \lim_{t\to\infty} |\tilde{\theta}|^2_{\Gamma^{-1}} - 2\beta \int_0^\infty \left[\frac{\partial V_a}{\partial t} \right.\\
&\quad + \frac{\partial V_a}{\partial e}\left(\tilde{f} + \tilde{F}\theta + F_r\tilde{\theta} - \frac{\partial \rho}{\partial \hat{\theta}}\Gamma\left(\frac{\partial V_a}{\partial e}F\right)^{\mathrm{T}} + g\tilde{u} \right) \\
&\quad \left. + \left(\frac{\partial V_a}{\partial \hat{\theta}} - \tilde{\theta}^{\mathrm{T}}\Gamma^{-1} \right)\Gamma\left(\frac{\partial V_a}{\partial e}F\right)^{\mathrm{T}} \right] dt + \int_0^\infty rv^2 dt \\
&= \beta \lim_{t\to\infty} |\tilde{\theta}|^2_{\Gamma^{-1}} - 2\beta \int_0^\infty d\left(V_a + \frac{1}{2}\tilde{\theta}^{\mathrm{T}}\Gamma^{-1}\tilde{\theta} \right) + \int_0^\infty rv^2 dt \\
&= 2\beta V_a(0, e(0), \hat{\theta}(0)) + \beta |\tilde{\theta}(0)|^2_{\Gamma^{-1}} \\
&\quad -2\beta \lim_{t\to\infty} V_a(t, e(t), \hat{\theta}(t)) + \int_0^\infty rv^2 dt. \quad (5.57)
\end{aligned}$$

Again, since $\tilde{u}(t)$ solves the adaptive quadratic tracking problem, $\lim_{t\to\infty} e(t) = 0$, and we have that $\lim_{t\to\infty} V_a(t, e(t), \hat{\theta}(t)) = 0$. Therefore, the minimum of (5.57) is reached only if $v = 0$, thus the control $\tilde{u} = \tilde{\alpha}^*(t, e, \hat{\theta})$ minimizes the cost functional (5.45). □

Remark 5.14 Even though not explicit in the proof of the above theorem, the atclf $V_a(t, e, \theta)$ solves the following family of Hamilton-Jacobi-Bellman equations parametrized in $\beta \geq 2$:

$$\frac{\partial V_a}{\partial t} + \frac{\partial V_a}{\partial e}\left[\tilde{f} + \tilde{F}\theta + F\Gamma\left(\frac{\partial V_a}{\partial \theta}\right)^{\mathrm{T}} - \underbrace{\frac{\partial \rho}{\partial \theta}\Gamma\left(\frac{\partial V_a}{\partial e}F\right)^{\mathrm{T}}}_{} \right] \\
- \frac{\beta}{2r(t, e, \theta)}\left(\frac{\partial V_a}{\partial e}g\right)^2 + \frac{l(t, e, \theta)}{2\beta} = 0. \quad (5.58)$$

The under-braced terms represent the "non-certainty-equivalence" part of this HJB equation. Their role is to take into account the time-varying effect of parameter adaptation and make the control law optimal *in the presence of an update law*. □

5.4 INVERSE OPTIMAL ADAPTIVE TRACKING

Remark 5.15 The freedom in selecting the parameter $\beta \geq 2$ in the control law (5.48) means that the inverse optimal adaptive controller has an infinite gain margin. □

Example 5.16 Consider the scalar *linear* system

$$\dot{x} = u + \theta x$$
$$y = x.$$
(5.59)

For simplicity, we focus on the regulation case, $y_r(t) \equiv 0$. Since the system is scalar, $V_a = \frac{1}{2}x^2$, $L_g V_a = x$, $L_f V_a = x^2 \theta$. We choose the control law based on Sontag's formula

$$u_s = -x\left(\theta + \sqrt{\theta^2 + 1}\right) = -2r^{-1}(\theta)x,$$
(5.60)

where

$$r(\theta) = \frac{2}{\theta + \sqrt{\theta^2 + 1}} > 0, \quad \forall \theta.$$
(5.61)

The control $\frac{u_s}{2}$ is stabilizing for the system (5.59) because

$$\dot{V}_a|_{\frac{u_s}{2}} = -\frac{1}{2}\left(-\theta + \sqrt{\theta^2 + 1}\right)x^2.$$
(5.62)

By Theorem 5.13, the control u_s is optimal with respect to the cost functional

$$J_a = \int_0^\infty \left(l(x,\theta) + ru^2\right) dt = 2\int_0^\infty \frac{x^2 + u^2}{\theta + \sqrt{\theta^2 + 1}} dt$$
(5.63)

with a value function $J_a^*(x) = 2x^2$. Meanwhile, the dynamic control

$$u_s = -x\left(\hat{\theta} + \sqrt{\hat{\theta}^2 + 1}\right)$$
(5.64)
$$\dot{\hat{\theta}} = x^2,$$
(5.65)

is optimal with respect to the cost functional

$$J = 2\left(\theta - \hat{\theta}(\infty)\right)^2 + 2\int_0^\infty \frac{x^2 + u^2}{\hat{\theta} + \sqrt{\hat{\theta}^2 + 1}} dt$$
(5.66)

with a value function $J^*(x, \hat{\theta}) = \left[x^2 + \left(\theta - \hat{\theta}\right)^2\right]$.

We point out that $\hat{\theta}(\infty)$ exists both due to the scalar (in parameter θ) nature of the problem and because it is a problem of regulation (see Chapter 7). Note that, even though the penalty coefficient on x and u in (5.66) varies with $\hat{\theta}(t)$, the penalty coefficient is always positive and finite. □

Remark 5.17 The control law (5.60) is, in fact, a linear-quadratic-regulator (LQR) for the system (5.59) when the parameter θ is known. The control law can be also written as $u_s = -p(\theta)x$ where $p(\theta)$ is the solution of the Riccati equation

$$p^2 - 2\theta p - 1 = 0. \tag{5.67}$$

It is of interest to compare the approach here with 'adaptive LQR schemes' for linear systems in [42, Section 7.4.4].

- Even though both methodologies result in the same control law (5.60) for the *scalar linear* system in Example 5.16, they employ different update laws. The *gradient* update law in [42, Section 7.4.4] is optimal with respect to an (instantaneous) cost on an *estimation* error, however, when its estimates $\hat{\theta}(t)$ are substituted into the control law (5.64), this control law is not (guaranteed to be) optimal for the overall system. (Even its proof of stability is a non-trivial matter!) In contrast, the update law (5.65) guarantees optimality of the control law (5.64) for the overall system with respect to the meaningful cost (5.66).

- The true difference between the approach here and the adaptive LQR scheme in [42, Section 7.4.4] arises for systems of higher order. Then the under-braced non-certainty-equivalence terms in (5.58) start to play a significant role. The CE approach in [42, Section 7.4.4] would be to set $\Gamma = 0$ in the HJB (Riccati—for linear systems) equation (5.58) and combine the resulting control law with a gradient or least-squares update law. The optimality of the non-adaptive controller would be lost in the presence of adaptation due to the time-varying $\hat{\theta}(t)$. In contrast, a solution to (5.14) with $\Gamma > 0$ would lead to optimality with respect to (5.45). □

Corollary 5.18 *If there exists an atclf $V_a(t, e, \theta)$ for (5.1) then the inverse optimal adaptive tracking problem is solvable.*

Proof. Consider the Sontag-type control law $u_s = \tilde{\alpha}(t, e, \theta)$ where $\tilde{\alpha}(t, e, \theta)$ is defined by (5.15). The control law $\frac{u_s}{2} = \frac{1}{2}\tilde{\alpha}(t, e, \theta)$ is an asymptotic stabilizing controller for system (5.9) because inequality (5.14) is satisfied with

$$W(t, e, \theta) = \frac{1}{2}\left[-\left(\frac{\partial V_a}{\partial t} + \frac{\partial V_a}{\partial e}\bar{f}\right) + \sqrt{\left(\frac{\partial V_a}{\partial t} + \frac{\partial V_a}{\partial e}\bar{f}\right)^2 + \left(\frac{\partial V_a}{\partial e}g\right)^4}\right], \tag{5.68}$$

which is positive definite in e (uniformly in t) for each θ by Lemma 5.4. Since $\frac{1}{2}\tilde{\alpha}(t,e,\theta)$ is of the form $\frac{1}{2}\tilde{\alpha}(t,e,\theta) = -r^{-1}\frac{\partial V_a}{\partial e}g$ with $r(t,e,\theta) > 0$ given by

$$r(t,e,\theta) = \begin{cases} \dfrac{2\left(\frac{\partial V_a}{\partial e}g\right)^2}{\frac{\partial V_a}{\partial t} + \frac{\partial V_a}{\partial e}\bar{f} + \sqrt{\left(\frac{\partial V_a}{\partial t} + \frac{\partial V_a}{\partial e}\bar{f}\right)^2 + \left(\frac{\partial V_a}{\partial e}g\right)^4}}, & \frac{\partial V_a}{\partial e}g \neq 0 \\ \text{any positive real number}, & \frac{\partial V_a}{\partial e}g = 0, \end{cases} \quad (5.69)$$

by Theorem 5.13, the inverse optimal adaptive tracking problem is solvable. The optimal control is the formula (5.15) itself. □

Corollary 5.19 *The inverse optimal adaptive tracking problem for the following system is solvable*

$$\begin{aligned} \dot{x}_i &= x_{i+1} + \varphi_i(x_1,\ldots,x_i)^T\theta, \quad i = 1,\ldots,n-1 \\ \dot{x}_n &= u + \varphi_n(x_1,\ldots,x_n)^T\theta \\ y &= x_1. \end{aligned} \quad (5.70)$$

Proof. By Corollary 5.11 and Corollary 5.8, there exists an atclf for (5.70). It then follows from Corollary 5.18 that the inverse optimal adaptive tracking problem for system (5.70) is solvable. □

5.5 Inverse Optimality via Backstepping

With Theorem 5.13, the problem of inverse optimal adaptive tracking is reduced to the problem of finding an atclf. However, the control law (5.15) based on Sontag's formula is not guaranteed to be smooth at the origin. In this section we develop controllers based on backstepping which are *smooth everywhere*, and, hence, can be employed in a recursive design.

Lemma 5.20 *If the adaptive quadratic tracking problem for the system*

$$\begin{aligned} \dot{x} &= f(x) + F(x)\theta + g(x)u \\ y &= h(x), \end{aligned} \quad (5.71)$$

is solvable with a smooth control law $\tilde{\alpha}(t,e,\theta)$ and (5.14) is satisfied with $W(t,e,\theta) = e^T\Omega(t,e,\theta)e$, where $\Omega(t,e,\theta)$ is positive definite and symmetric for all t,e,θ; then the inverse optimal adaptive tracking problem for the augmented system

$$\begin{aligned} \dot{x} &= f(x) + F(x)\theta + g(x)\xi \\ \dot{\xi} &= u \\ y &= h(x), \end{aligned} \quad (5.72)$$

is also solvable with a smooth *control law.*

Proof. Since the adaptive quadratic tracking problem for the system (5.71) is solvable, by Lemma 5.9 and Corollary 5.8, $V_1(t,e,\tilde{\xi},\theta) = V_a(t,e,\theta) + \frac{1}{2}(\tilde{\xi} - \tilde{\alpha}(t,e,\theta))^2$ is an atclf for the augmented system (5.72), i.e., a clf for the modified nonadaptive error system (5.35). Adding (5.38) and (5.39), with (5.14), we get

$$\begin{aligned}
\dot{V}_1 &= \frac{\partial V_1}{\partial t} + \frac{\partial V_1}{\partial (e,\tilde{\xi})} \begin{bmatrix} \tilde{f} + \tilde{F}\theta + F\Gamma \left(\frac{\partial V_1}{\partial \theta}\right)^T - \frac{\partial \rho}{\partial \theta}\Gamma \left(\frac{\partial V_1}{\partial e}F\right)^T + g\tilde{\xi} \\ \tilde{u} - \frac{\partial \alpha_r}{\partial \theta}\Gamma \left(\frac{\partial V_1}{\partial e}F\right)^T \end{bmatrix} \\
&\leq -W + z\left[\tilde{u} - \frac{\partial \tilde{\alpha}}{\partial t} + \frac{\partial V_a}{\partial e}g - \frac{\partial \tilde{\alpha}}{\partial e}\left(\tilde{f} + \tilde{F}\theta + g(\tilde{\alpha} + z)\right)\right. \\
&\quad - \left(\frac{\partial \alpha_r}{\partial \theta} + \frac{\partial \tilde{\alpha}}{\partial \theta} - \frac{\partial \tilde{\alpha}}{\partial e}\frac{\partial \rho}{\partial \theta}\right)\Gamma \left(\frac{\partial V_a}{\partial e}F\right)^T \\
&\quad + \left(\frac{\partial \alpha_r}{\partial \theta} + \frac{\partial \tilde{\alpha}}{\partial \theta} - \frac{\partial \tilde{\alpha}}{\partial e}\frac{\partial \rho}{\partial \theta}\right)\Gamma \left(\frac{\partial \tilde{\alpha}}{\partial e}F\right)^T z \\
&\quad \left. - \left(\frac{\partial V_a}{\partial \theta} - \frac{\partial V_a}{\partial e}\frac{\partial \rho}{\partial \theta}\right)\Gamma \left(\frac{\partial \tilde{\alpha}}{\partial e}F\right)^T\right],
\end{aligned} \qquad (5.73)$$

where $z = \tilde{\xi} - \tilde{\alpha}(t,e,\theta)$. To render $\dot{V}_1$ negative definite, one choice is (5.36) which cancels all the nonlinear terms inside the bracket in (5.73). However, the cancellation controller (5.36) is not (guaranteed to be) optimal. Therefore, we have to use other techniques in the design of our control law. One such technique we will use here is "nonlinear damping."

Since $\tilde{\alpha}$, $\frac{\partial \tilde{\alpha}}{\partial t}$, $\tilde{f}$, $\tilde{F}$, $\frac{\partial V_a}{\partial e}$, $\frac{\partial V_a}{\partial \theta}$ are smooth and vanish for $e = 0$, then we can write

$$\begin{aligned}
&-\frac{\partial \tilde{\alpha}}{\partial t} + \frac{\partial V_a}{\partial e}g - \frac{\partial \tilde{\alpha}}{\partial e}\left(\tilde{f} + \tilde{F}\theta + g\tilde{\alpha}\right) \\
&- \left(\frac{\partial \alpha_r}{\partial \theta} + \frac{\partial \tilde{\alpha}}{\partial \theta} - \frac{\partial \tilde{\alpha}}{\partial e}\frac{\partial \rho}{\partial \theta}\right)\Gamma \left(\frac{\partial V_a}{\partial e}F\right)^T - \left(\frac{\partial V_a}{\partial \theta} - \frac{\partial V_a}{\partial e}\frac{\partial \rho}{\partial \theta}\right)\Gamma \left(\frac{\partial \tilde{\alpha}}{\partial e}F\right)^T \\
&= \Psi_1(t,e,\theta)^T \Omega(t,e,\theta)^{1/2} e,
\end{aligned} \qquad (5.74)$$

where $\Psi_1(t,e,\theta)$ is a vector-valued smooth function and $\Omega(t,e,\theta)^{1/2}$ is invertible for all t,e,θ. In addition, let us denote

$$-\frac{\partial \tilde{\alpha}}{\partial e}g + \left(\frac{\partial \alpha_r}{\partial \theta} + \frac{\partial \tilde{\alpha}}{\partial \theta} - \frac{\partial \tilde{\alpha}}{\partial e}\frac{\partial \rho}{\partial \theta}\right)\Gamma \left(\frac{\partial \tilde{\alpha}}{\partial e}F\right)^T = \Psi_2(t,e,\theta). \qquad (5.75)$$

Then (5.73) is re-written as

$$\dot{V}_1 \leq -|\Omega^{1/2}e|^2 + z\tilde{u} + z\Psi_1\Omega^{1/2}e + \Psi_2 z^2. \qquad (5.76)$$

5.5 INVERSE OPTIMALITY VIA BACKSTEPPING

The choice

$$\tilde{u} = \tilde{\alpha}_1(t, e, \tilde{\xi}, \theta) = -\left(c + \frac{|\Psi_1|^2}{2} + \frac{\Psi_2^2}{2c}\right)z, \quad c > 0, \tag{5.77}$$

renders

$$\dot{V}_1 \leq -\frac{1}{2}|\Omega^{1/2}e|^2 - \frac{c}{2}z^2. \tag{5.78}$$

Since the control law $\tilde{u} = \tilde{\alpha}_1$ defined in (5.77) is of the form

$$\tilde{\alpha}_1(t, e, \tilde{\xi}, \theta) = -R^{-1}(t, e, \tilde{\xi}, \theta)\frac{\partial V_1}{\partial(e, \tilde{\xi})}\begin{bmatrix} 0 \\ 1 \end{bmatrix}, \tag{5.79}$$

where

$$R^{-1}(t, e, \tilde{\xi}, \theta) = \left(c + \frac{|\Psi_1|^2}{2} + \frac{\Psi_2^2}{2c}\right) > 0, \quad \forall t, e, \tilde{\xi}, \theta, \tag{5.80}$$

by Theorem 5.13, the dynamic feedback control $\tilde{u}^* = \beta\tilde{\alpha}_1(t, e, \tilde{\xi}, \hat{\theta})$, $\beta \geq 2$, with

$$\dot{\hat{\theta}} = \Gamma\tau_1(t, e, \tilde{\xi}, \hat{\theta}) = \Gamma\left(\frac{\partial V_1}{\partial e}F(t, e, \tilde{\xi}, \hat{\theta})\right)^T, \tag{5.81}$$

is optimal for the closed-loop tracking error system (5.33) and (5.81). □

Example 5.21 (Example 5.10, revisited) For the system (5.41), we designed a controller (5.42) which is not optimal due to its cancellation property. With Lemma 5.20, we can design an optimal control as follows. First we note that $\tilde{\alpha}$ given by (5.27) in Example 5.6 is of the form

$$\begin{aligned}\tilde{\alpha}(t, e, \theta) &= -\left[1 + \varphi^T\Gamma\varphi - \left(c_1 + \frac{\partial\tilde{\varphi}^T}{\partial e_1}\theta\right)c_1 + \frac{\partial\phi^T}{\partial t}\theta\right]z_1 \\ &\quad - \left[c_2 + \left(c_1 + \frac{\partial\tilde{\varphi}^T}{\partial e_1}\theta\right)(1 + \varphi^T\Gamma\varphi)\right]z_2 \\ &:= a(t, e, \theta)z_1 + b(t, e, \theta)z_2, \end{aligned} \tag{5.82}$$

because $\tilde{\varphi} = \varphi(x_1) - \varphi(x_{r1}) = \varphi(e_1 + x_{r1}) - \varphi(x_{r1}) = e_1\phi(e_1)$ and $\frac{\partial\tilde{\varphi}}{\partial t} = z_1\frac{\partial\phi}{\partial t}$. Instead of (5.42) we choose the "nonlinear damping" control suggested by Lemma 5.20:

$$\begin{aligned}\tilde{u} &= \tilde{\alpha}_1(t, e, \tilde{\xi}, \theta) \\ &= -\left\{c_3 + \frac{1}{2c_1}\left[\frac{\partial\tilde{\alpha}}{\partial e_1}c_1 - \frac{\partial a}{\partial t} - \frac{\partial\tilde{\alpha}}{\partial e_2}a - \left(\frac{\partial\alpha_r}{\partial\theta} + \frac{\partial\tilde{\alpha}}{\partial\theta} + \frac{\partial\tilde{\alpha}}{\partial e_2}\varphi_r^T\right)\Gamma\varphi\right]^2 \\ &\quad + \frac{1}{2c_2}\left[1 + \frac{\partial\tilde{\alpha}}{\partial e_1}\varphi^T\Gamma\varphi - \frac{\partial\tilde{\alpha}}{\partial e_1} - \frac{\partial b}{\partial t} - \frac{\partial\tilde{\alpha}}{\partial e_2}b \right.\\ &\quad \left. - \left(\frac{\partial\alpha_r}{\partial\theta} + \frac{\partial\tilde{\alpha}}{\partial\theta} + \frac{\partial\tilde{\alpha}}{\partial e_2}\varphi_r^T\right)\Gamma\varphi\left(c_1 + \frac{\partial\tilde{\varphi}^T}{\partial e_1}\theta\right)\right]^2 + \frac{1}{2c_3}\left(\frac{\partial\tilde{\alpha}}{\partial e_2}\right)^2\right\}z_3. \end{aligned} \tag{5.83}$$

The tuning function τ_1 is the same as in (5.43). The control law $\tilde{u}^* = \beta\tilde{\alpha}_1(t,e,\tilde{\xi},\hat{\theta})$, $\beta \geq 2$, with $\dot{\hat{\theta}} = \Gamma\tau_1$ is optimal. □

5.6 Design for Strict-Feedback Systems

We now consider the *parametric strict-feedback systems*:

$$\begin{aligned}
\dot{x}_i &= x_{i+1} + \varphi_i(\bar{x}_i)^\mathrm{T}\theta, \quad i = 1,\ldots,n-1 \\
\dot{x}_n &= u + \varphi_n(x)^\mathrm{T}\theta \\
y &= x_1
\end{aligned} \tag{5.84}$$

and develop a procedure for optimal adaptive tracking of a given signal $y_r(t)$. An inverse optimal design following from Corollary 5.19 (based on Sontag's formula) would be non-smooth at $e = 0$. In this section we develop a design which is smooth everywhere.

For the class of systems (5.84), Assumption 5.1 is satisfied for any function $y_r(t)$, and there exist functions $\rho_1(t)$, $\rho_2(t,\theta)$, $\cdots$, $\rho_n(t,\theta)$, and $\alpha_r(t,\theta)$ such that

$$\begin{aligned}
\dot{\rho}_i &= \rho_{i+1} + \varphi_i(\bar{\rho}_i)^\mathrm{T}\theta, \quad i = 1,\ldots,n-1 \\
\dot{\rho}_n &= \alpha_r(t,\theta) + \varphi_n(\rho)^\mathrm{T}\theta \\
y_r(t) &= \rho_1(t).
\end{aligned} \tag{5.85}$$

Consider the signal $x_r(t) = \rho(t,\hat{\theta})$ which is governed by

$$\begin{aligned}
\dot{x}_{ri} &= x_{r,i+1} + \varphi_{ri}(\bar{x}_{ri})^\mathrm{T}\hat{\theta} + \frac{\partial\rho_i}{\partial\hat{\theta}}\dot{\hat{\theta}}, \quad i = 1,\ldots,n-1 \\
\dot{x}_{rn} &= \alpha_r(t,\hat{\theta}) + \varphi_{rn}(x_r)^\mathrm{T}\hat{\theta} + \frac{\partial\rho_n}{\partial\hat{\theta}}\dot{\hat{\theta}} \\
y_r &= x_{r1}.
\end{aligned} \tag{5.86}$$

The tracking error $e = x - x_r$ is governed by the system

$$\begin{aligned}
\dot{e}_i &= e_{i+1} + \tilde{\varphi}_i(t,\bar{e}_i,\hat{\theta})^\mathrm{T}\theta + \varphi_{ri}(t,\hat{\theta})^\mathrm{T}\tilde{\theta} - \frac{\partial\rho_i}{\partial\hat{\theta}}\dot{\hat{\theta}}, \quad i = 1,\ldots,n-1 \\
\dot{e}_n &= \tilde{u} + \tilde{\varphi}_n(t,e,\hat{\theta})^\mathrm{T}\theta + \varphi_{rn}(t,\hat{\theta})^\mathrm{T}\tilde{\theta} - \frac{\partial\rho_n}{\partial\hat{\theta}}\dot{\hat{\theta}}.
\end{aligned}$$
(5.87)

where $\tilde{u} = u - \alpha_r(t,\hat{\theta})$ and $\tilde{\varphi}_i = \varphi_i(\bar{x}_i) - \varphi_{ri}(\bar{x}_{ri})$, $i =,\cdots,n$. For an atclf V_a,

5.6 DESIGN FOR STRICT-FEEDBACK SYSTEMS

the modified non-adaptive error system is

$$\begin{aligned}
\dot{e}_i &= e_{i+1} + \tilde{\varphi}_i^{\mathrm{T}}\theta + \varphi_i^{\mathrm{T}}\Gamma\left(\frac{\partial V_a}{\partial \theta}\right)^{\mathrm{T}} - \frac{\partial \rho_i}{\partial \hat{\theta}}\Gamma\left(\frac{\partial V_a}{\partial e}F\right)^{\mathrm{T}}, \quad i = 1,\ldots,n-1 \\
\dot{e}_n &= \tilde{u} + \tilde{\varphi}_n(t,e,\hat{\theta})^{\mathrm{T}}\theta + \varphi_n^{\mathrm{T}}\Gamma\left(\frac{\partial V_a}{\partial \theta}\right)^{\mathrm{T}} - \frac{\partial \rho_n}{\partial \hat{\theta}}\Gamma\left(\frac{\partial V_a}{\partial e}F\right)^{\mathrm{T}},
\end{aligned} \quad (5.88)$$

where $F = [\varphi_1^{\mathrm{T}},\cdots,\varphi_n^{\mathrm{T}}]^{\mathrm{T}}$.

First, we search for an atclf for the system (5.84). Repeated application of Lemma 5.9 gives an atclf

$$\begin{aligned}
V_a &= \frac{1}{2}\sum_{i=1}^{n} z_i^2 \\
z_i &= e_i - \tilde{\alpha}_{i-1}(t,\bar{e}_{i-1},\theta),
\end{aligned} \quad (5.89)$$

where $\tilde{\alpha}_i$'s are to be determined. For notational convenience we define $z_0 := 0$, $\tilde{\alpha}_0 := 0$. We then have

$$\frac{\partial V_a}{\partial \theta} = -\sum_{j=1}^{n} \frac{\partial \tilde{\alpha}_{j-1}}{\partial \theta} z_j \quad (5.90)$$

$$\left(\frac{\partial V_a}{\partial e}F\right)^{\mathrm{T}} = \sum_{j=1}^{n} \frac{\partial V_a}{\partial e_j}\varphi_j = \sum_{j=1}^{n}\left(z_j - \sum_{k=j+1}^{n}\frac{\partial \tilde{\alpha}_{k-1}}{\partial e_j}z_k\right)\varphi_j = \sum_{j=1}^{n} w_j z_j, \quad (5.91)$$

where

$$w_j(t,\bar{e}_j,\theta) = \varphi_j - \sum_{k=1}^{j-1}\frac{\partial \tilde{\alpha}_{j-1}}{\partial e_k}\varphi_k. \quad (5.92)$$

Therefore the modified non-adaptive error system (5.88) becomes

$$\begin{aligned}
\dot{e}_i &= e_{i+1} + \tilde{\varphi}_i^{\mathrm{T}}\theta - \sum_{j=1}^{n}\frac{\partial \tilde{\alpha}_{j-1}}{\partial \theta}\Gamma\varphi_i z_j - \sum_{j=1}^{n}\frac{\partial \rho_i}{\partial \hat{\theta}}\Gamma w_j z_j, \quad i=1,\ldots,n-1 \\
\dot{e}_n &= \tilde{u} + \tilde{\varphi}_n^{\mathrm{T}}\theta - \sum_{j=1}^{n}\frac{\partial \tilde{\alpha}_{j-1}}{\partial \theta}\Gamma\varphi_n z_j - \sum_{j=1}^{n}\frac{\partial \rho_n}{\partial \hat{\theta}}\Gamma w_j z_j.
\end{aligned} \quad (5.93)$$

The functions $\tilde{\alpha}_1, \cdots, \tilde{\alpha}_{n-1}$ are yet to be determined to make V_a defined in (5.89) a clf for system (5.93). To design these functions, we apply the backstepping technique. We perform cancellations at all the steps before step n. At the final step n, we choose the actual control $\tilde{u}$ in a form which, according to Lemma 5.20, is inverse optimal.

Step $i = 1,\cdots,n-1$:

$$\tilde{\alpha}_i(t,\bar{e}_i,\theta) = -z_{i-1} - c_i z_i + \frac{\partial \tilde{\alpha}_{i-1}}{\partial t} + \sum_{k=1}^{i-1}\frac{\partial \tilde{\alpha}_{i-1}}{\partial e_k}e_{k+1} - \tilde{w}_i^{\mathrm{T}}\theta$$

$$-\sum_{k=1}^{i-1}(\sigma_{ki}+\sigma_{ik})z_k - \sigma_{ii}z_i, \quad c_i > 0 \tag{5.94}$$

$$\tilde{w}_i(t, \bar{e}_i, \theta) = \tilde{\varphi}_i - \sum_{k=1}^{i-1}\frac{\partial \tilde{\alpha}_{i-1}}{\partial e_k}\tilde{\varphi}_k \tag{5.95}$$

$$\sigma_{ik} = -\left(\frac{\partial \tilde{\alpha}_{i-1}}{\partial \theta} + \frac{\partial \rho_i}{\partial \theta} - \sum_{j=2}^{i-1}\frac{\partial \tilde{\alpha}_{i-1}}{\partial e_j}\frac{\partial \rho_j}{\partial \theta}\right)\Gamma w_k. \tag{5.96}$$

Before we go to the last step, we present the following lemma.

Lemma 5.22 *The time derivative of V_a in (5.89) along the solutions of system (5.93) with (5.94)—(5.96) is given by*

$$\dot{V}_a = -\sum_{k=1}^{n-1} c_k z_k^2 + z_n \left[z_{n-1} + \sum_{k=1}^{n-1}(\sigma_{kn}+\sigma_{nk})z_k + \sigma_{nn}z_n + \tilde{u} \right.$$
$$\left. - \frac{\partial \tilde{\alpha}_{n-1}}{\partial t} - \sum_{k=1}^{n-1}\frac{\partial \tilde{\alpha}_{n-1}}{\partial e_k}e_{k+1} + \tilde{w}_n^T\theta \right]. \tag{5.97}$$

Proof. First we prove that the closed-loop system after i steps is

$$\begin{bmatrix} \dot{z}_1 \\ \vdots \\ \dot{z}_i \end{bmatrix} = \begin{bmatrix} -c_1 & 1+\pi_{12} & \pi_{13} & \cdots & \pi_{1,i-1} & \pi_{1i} \\ -1-\pi_{12} & -c_2 & 1+\pi_{23} & \cdots & \pi_{2,i-1} & \pi_{2i} \\ -\pi_{13} & -1-\pi_{23} & \ddots & \ddots & \vdots & \vdots \\ \vdots & \vdots & \ddots & \ddots & 1+\pi_{i-2,i-1} & \pi_{i-2,i} \\ -\pi_{1,i-1} & -\pi_{2,i-1} & \cdots & -1-\pi_{i-2,i-1} & -c_{i-1} & 1+\pi_{i-1,i} \\ -\pi_{1i} & -\pi_{2i} & \cdots & -\pi_{i-2,i} & -1-\pi_{i-1,i} & -c_i \end{bmatrix}$$

$$\times \begin{bmatrix} z_1 \\ \vdots \\ z_i \end{bmatrix} + \begin{bmatrix} \pi_{1,i+1} & \pi_{1,i+2} & \cdots & \pi_{1n} \\ \pi_{2,i+1} & \pi_{2,i+2} & \cdots & \pi_{2n} \\ \vdots & & & \vdots \\ \pi_{i-1,i+1} & \pi_{i-1,i+2} & \cdots & \pi_{i-1,n} \\ 1+\pi_{i,i+1} & \pi_{i,i+2} & \cdots & \pi_{i,n} \end{bmatrix} \begin{bmatrix} z_{i+1} \\ \vdots \\ z_n \end{bmatrix}, \tag{5.98}$$

and the resulting $\dot{V}_i$ is

$$\dot{V}_i = -\sum_{k=1}^{i} c_k z_k^2 + z_i z_{i+1} + \sum_{j=i+1}^{n} z_j \sum_{k=1}^{i} \pi_{kj} z_k, \tag{5.99}$$

where

$$\pi_{ik} = \eta_{ik} + \xi_{ik} \tag{5.100}$$

$$\eta_{ik} = -\frac{\partial \tilde{\alpha}_{k-1}}{\partial \theta}\Gamma w_i \tag{5.101}$$

$$\xi_{ik} = -\left(\frac{\partial \rho_i}{\partial \theta} - \sum_{j=2}^{i-1}\frac{\partial \tilde{\alpha}_{k-1}}{\partial e_j}\frac{\partial \rho_j}{\partial \theta}\right)\Gamma w_k. \tag{5.102}$$

5.6 DESIGN FOR STRICT-FEEDBACK SYSTEMS

The proof is by induction. Step 1: Substituting $\tilde{\alpha}_1 = -c_1 z_1 - \tilde{\varphi}_1^T \theta$ into (5.93) with $i = 1$, using (5.89) and noting $\tilde{\alpha}_0 = 0$ and $\frac{\partial \rho_1}{\partial \theta} = 0$, we get

$$\dot{z}_1 = -c_1 z_1 + z_2 - \sum_{j=1}^{n} z_j \frac{\partial \tilde{\alpha}_{j-1}}{\partial \theta} \Gamma w_1 = -c_1 z_1 + z_2 - \pi_{12} z_3 - \cdots - \pi_{1n} z_n, \quad (5.103)$$

and

$$\dot{V}_1 = -c_1 z_1^2 + z_1 z_2 - \sum_{j=1}^{n} z_j \frac{\partial \tilde{\alpha}_{j-1}}{\partial \theta} \Gamma w_1 z_1 = -c_1 z_1^2 + z_1 z_2 - \sum_{j=2}^{n} z_j \pi_{1j} z_1, \quad (5.104)$$

which shows that (5.98) and (5.99) are true for $i = 1$.

Assume that (5.98) and (5.99) are true for $i - 1$, that is,

$$\begin{bmatrix} \dot{z}_1 \\ \vdots \\ \dot{z}_{i-1} \end{bmatrix} = \begin{bmatrix} -c_1 & 1+\pi_{12} & \pi_{13} & \cdots & \pi_{1,i-2} & \pi_{1,i-1} \\ -1-\pi_{12} & -c_2 & 1+\pi_{23} & \cdots & \pi_{2,i-2} & \pi_{2,i-1} \\ -\pi_{13} & -1-\pi_{23} & \ddots & \ddots & \vdots & \vdots \\ \vdots & \vdots & \ddots & \ddots & 1+\pi_{i-2,i-2} & \pi_{i-2,i-1} \\ -\pi_{1,i-2} & -\pi_{2,i-2} & \cdots & -1-\pi_{i-3,i-2} & -c_{i-2} & 1+\pi_{i-2,i-1} \\ -\pi_{1,i-1} & -\pi_{2,i-1} & \cdots & -\pi_{i-3,i-1} & -1-\pi_{i-2,i-1} & -c_{i-1} \end{bmatrix}$$

$$\times \begin{bmatrix} z_1 \\ \vdots \\ z_{i-1} \end{bmatrix} + \begin{bmatrix} \pi_{1,i} & \pi_{1,i+1} & \cdots & \pi_{1n} \\ \pi_{2,i} & \pi_{2,i+1} & \cdots & \pi_{2n} \\ \vdots & & & \\ \pi_{i-2,i} & \pi_{i-2,i+1} & \cdots & \pi_{i-2,n} \\ 1+\pi_{i-1,i} & \pi_{i-1,i+1} & \cdots & \pi_{i-1,n} \end{bmatrix} \begin{bmatrix} z_i \\ \vdots \\ z_n \end{bmatrix}, \quad (5.105)$$

and

$$\dot{V}_{i-1} = -\sum_{k=1}^{i-1} c_k z_k^2 + z_{i-1} z_i + \sum_{j=i}^{n} z_j \sum_{k=1}^{i-1} \pi_{kj} z_k. \quad (5.106)$$

The z_i-subsystem is

$$\dot{z}_i = z_{i+1} + \tilde{\alpha}_i + \tilde{\varphi}_i^T \theta - \sum_{j=1}^{n} \frac{\partial \tilde{\alpha}_{j-1}}{\partial \theta} \Gamma \varphi_i z_j - \sum_{j=1}^{n} \frac{\partial \rho_i}{\partial \theta} \Gamma w_j z_j$$

$$- \frac{\partial \tilde{\alpha}_{i-1}}{\partial t} - \sum_{k=1}^{i-1} \frac{\partial \tilde{\alpha}_{i-1}}{\partial e_k} \left(e_{k+1} + \tilde{\varphi}_k^T \theta - \sum_{j=1}^{n} \frac{\partial \tilde{\alpha}_{j-1}}{\partial \theta} \Gamma \varphi_k z_j - \sum_{j=1}^{n} \frac{\partial \rho_k}{\partial \theta} \Gamma w_j z_j \right)$$

$$= z_{i+1} + \tilde{\alpha}_i - \frac{\partial \tilde{\alpha}_{i-1}}{\partial t} - \sum_{k=1}^{i-1} \frac{\partial \tilde{\alpha}_{i-1}}{\partial e_k} e_{k+1} + \tilde{w}_i^T \theta + \sum_{j=1}^{n} \pi_{ij} z_j. \quad (5.107)$$

The derivative of $V_i = V_{i-1} + \frac{1}{2} z_i^2$ is calculated as

$$\dot{V}_i = -\sum_{k=1}^{i-1} c_k z_k^2 + z_i z_{i+1} + \sum_{j=i+1}^{n} z_j \sum_{k=1}^{i-1} \pi_{kj} z_k + \sum_{j=i+1}^{n} z_j \pi_{ij} z_i$$

$$+ z_i \left[z_{i-1} + \sum_{k=1}^{i-1} \pi_{ki} z_k + \tilde{\alpha}_i - \frac{\partial \tilde{\alpha}_{i-1}}{\partial t} - \sum_{k=1}^{i-1} \frac{\partial \tilde{\alpha}_{i-1}}{\partial e_k} e_{k+1} + \tilde{w}_i^T \theta + \sum_{k=1}^{i} \pi_{ik} z_k \right]$$

$$= -\sum_{k=1}^{i-1} c_k z_k^2 + z_i z_{i+1} + \sum_{j=i+1}^{n} z_j \sum_{k=1}^{i} \pi_{kj} z_k$$

$$+ z_i \left[\tilde{\alpha}_i + z_{i-1} - \frac{\partial \tilde{\alpha}_{i-1}}{\partial t} - \sum_{k=1}^{i-1} \frac{\partial \tilde{\alpha}_{i-1}}{\partial e_k} e_{k+1} + \tilde{w}_i^T \theta + \sum_{k=1}^{i-1} (\pi_{ki} + \pi_{ik}) z_k + \pi_{ii} z_i \right]. \tag{5.108}$$

From the definitions of π_{ik}, η_{ik}, δ_{ik} and σ_{ik}, it is easy to show that $\pi_{ki} + \pi_{ik} = \sigma_{ki} + \sigma_{ik}$ and that $\pi_{ii} = \sigma_{ii}$. Therefore the choice of $\tilde{\alpha}_i$ as in (5.94) results in (5.99) and

$$\dot{z}_i = -\sum_{k=1}^{i-2} \pi_{ki} z_k - (1 + \pi_{i-1,i}) z_{i-1} - c_i z_i + (1 + \pi_{i,i+1}) z_{i+1} + \sum_{k=i+2}^{n} \pi_{ik} z_k. \tag{5.109}$$

Combining (5.109) with (5.105), we get (5.98).

We now rewrite the last equation of (5.93) as

$$\begin{aligned}
\dot{z}_n &= \tilde{u} + \tilde{\varphi}_n^T \theta - \sum_{j=1}^{n} \frac{\partial \tilde{\alpha}_{j-1}}{\partial \theta} \Gamma \varphi_n z_j - \sum_{j=1}^{n} \frac{\partial \rho_n}{\partial \theta} \Gamma w_j z_j \\
&\quad - \frac{\partial \tilde{\alpha}_{n-1}}{\partial t} - \sum_{k=1}^{n-1} \frac{\partial \tilde{\alpha}_{n-1}}{\partial e_k} \left(e_{k+1} + \tilde{\varphi}_k^T \theta - \sum_{j=1}^{n} \frac{\partial \tilde{\alpha}_{j-1}}{\partial \theta} \Gamma \varphi_k z_j - \sum_{j=1}^{n} \frac{\partial \rho_k}{\partial \theta} \Gamma w_j z_j \right) \\
&= \tilde{u} - \frac{\partial \tilde{\alpha}_{n-1}}{\partial t} - \sum_{k=1}^{n-1} \frac{\partial \tilde{\alpha}_{n-1}}{\partial e_k} e_{k+1} + \tilde{w}_n^T \theta + \sum_{j=1}^{n} \pi_{nj} z_j.
\end{aligned} \tag{5.110}$$

where $\tilde{w}_n$ follows the same definition as in (5.95). Noting that $V_a = V_{n-1} + \frac{1}{2} z_n^2$ and $\pi_{kn} + \pi_{nk} = \sigma_{kn} + \sigma_{nk}$, and $\pi_{nn} = \sigma_{nn}$, (5.97) follows readily from (5.99) and (5.110). $\square$

Step n: With the help of Lemma 5.22, the derivative of V_a is (5.97). We are now at the position to choose the actual control $\tilde{u}$. We may choose $\tilde{u}$ such that all the terms inside the bracket are cancelled and the bracketed term multiplying z_n is equal to $-c_n z_n^2$, but the controller designed in that way is not guaranteed to be inverse optimal. To design a controller which is inverse optimal, according to Theorem 5.13, we should choose a control law that is of the form

$$\tilde{u} = \tilde{\alpha}_n(t, e, \theta) = -r^{-1}(t, e, \theta) \frac{\partial V_a}{\partial e} g, \tag{5.111}$$

where $r(t, e, \theta) > 0$, $\forall t, e, \theta$. In light of (5.93) and (5.89), (5.111) simplifies to

$$\tilde{u} = \tilde{\alpha}_n(t, e, \theta) = -r^{-1}(t, e, \theta) z_n, \tag{5.112}$$

i.e., we must choose $\tilde{\alpha}_n$ with z_n as a factor.

Since $e_{k+1} = z_{k+1} + \tilde{\alpha}_k$, $k = 1, \cdots, n-1$, and the expression in the second line in (5.97) vanishes at $e = 0$, it is easy to see that it also vanishes for $z = 0$. Therefore, there exist smooth functions ϕ_k, $k = 1, \cdots, n$, such that

$$-\frac{\partial \tilde{\alpha}_{n-1}}{\partial t} - \sum_{k=1}^{n-1} \frac{\partial \tilde{\alpha}_{n-1}}{\partial e_k} e_{k+1} + \tilde{w}_n^T \theta = \sum_{k=1}^{n} \phi_k z_k. \qquad (5.113)$$

Thus (5.97) becomes

$$\dot{V}_a = -\sum_{k=1}^{n-1} c_k z_k^2 + z_n \tilde{u} + \sum_{k=1}^{n} z_n \Phi_k z_k, \qquad (5.114)$$

where

$$\Phi_k = \sigma_{kn} + \sigma_{nk} + \phi_k, \quad k = 1, \cdots, n-2$$
$$\Phi_{n-1} = 1 + \sigma_{n-1,n} + \sigma_{n,n-1} + \phi_{n-1} \qquad (5.115)$$
$$\Phi_n = \sigma_{nn} + \phi_n.$$

A control law of the form (5.112) with

$$r(t, e, \theta) = \left(c_n + \sum_{k=1}^{n} \frac{\Phi_k^2}{2c_k}\right)^{-1} > 0, \; c_k > 0, \; \forall t, e, \theta, \qquad (5.116)$$

results in

$$\dot{V}_a = -\frac{1}{2} \sum_{k=1}^{n} c_k z_k^2 - \sum_{k=1}^{n} \frac{c_k}{2} \left(z_k - \frac{\Phi_k}{c_k} z_n\right)^2. \qquad (5.117)$$

By Theorem 5.13, the inverse optimal adaptive tracking problem is solved through the dynamic feedback control (adaptive control) law

$$\tilde{u} = \tilde{\alpha}_n^*(t, e, \hat{\theta}) = 2\tilde{\alpha}_n(t, e, \hat{\theta})$$
$$\dot{\hat{\theta}} = \Gamma \left(\frac{\partial V_a}{\partial e} F\right)^T = \Gamma \sum_{j=1}^{n} w_j z_j. \qquad (5.118)$$

5.7 Transient Performance

In this brief section, we give an $\mathcal{L}_2$ bound on the error state z and control $\tilde{u}$ for the inverse optimal adaptive controller designed in Section 5.6. According to Theorem 5.13, the control law (5.118) is optimal with respect to the cost functional

$$J = 2 \lim_{t \to \infty} |\theta - \hat{\theta}(t)|^2_{\Gamma^{-1}}$$
$$+ 2 \int_0^\infty \left[\sum_{k=1}^{n} c_k z_k^2 + \sum_{k=1}^{n} c_k \left(z_k - \frac{\Phi_k}{c_k} z_n\right)^2 + \frac{\tilde{u}^2}{2\left(c_n + \sum_{k=1}^{n} \frac{\Phi_k^2}{2c_k}\right)}\right] dt \qquad (5.119)$$

with a value function

$$J^* = 2|\theta - \hat{\theta}|_{\Gamma^{-1}}^2 + 2|z|^2. \tag{5.120}$$

In particular, we have the following $\mathcal{L}_2$ performance result.

Theorem 5.23 *In the adaptive system (5.87), (5.118), the following inequality holds*

$$\int_0^\infty \left[\sum_{k=1}^n c_k z_k^2 + \frac{\tilde{u}^2}{2\left(c_n + \sum_{k=1}^n \frac{\Phi_k^2}{2c_k}\right)} \right] dt \leq |\tilde{\theta}(0)|_{\Gamma^{-1}}^2 + |z(0)|^2. \tag{5.121}$$

The bound (5.121) depends on $z(0)$ which is dependent on the design parameters $c_1, \cdots, c_n$, and Γ. To eliminate this dependency and allow a systematic improvement of the bound on $\|z\|_2$, we employ *trajectory initialization* as in [63, Section 4.3.2] to set $z(0) = 0$ and obtain:

$$\int_0^\infty \left[\sum_{k=1}^n c_k z_k^2 + \frac{\tilde{u}^2}{2\left(c_n + \sum_{k=1}^n \frac{\Phi_k^2}{2c_k}\right)} \right] dt \leq |\tilde{\theta}(0)|_{\Gamma^{-1}}^2. \tag{5.122}$$

Notes and References

A thorough presentation of adaptive control for nonlinear systems is given in the book by Krstić, Kanellakopoulos and Kokotović [63]. The many methods presented in [63] are outgrowth of Kanellakopoulos' adaptive backstepping [55]. The reader's attention is also drawn to the book by Marino and Tomei [77] which specializes in differential geometric aspects and output feedback. The book of Tao and Kokotović [104] develops adaptive controllers for systems with actuator and sensor nonlinearities. While [63, 77] solve only stabilization/tracking problems, this chapter, based on Li and Krstić [70], solves also an *optimal* control problem.

The optimal adaptive control problem posed here is not entirely dissimilar from the problem posed by Didinsky and Başar [17] and solved using their cost-to-come method. The difference is twofold: (a) the approach here does not require the inclusion of a noise term in the plant model in order to be able to design a parameter estimator, (b) while [17] only goes as far as to derive a Hamilton-Jacobi-Isaacs equation whose solution would yield an optimal controller, the HJB equation is actually solved here for strict-feedback systems. A nice marriage of the work of [17] and the backstepping design in [63] was brought out in the paper by Pan and Başar [88] who solved an adaptive disturbance attenuation problem for strict-feedback systems. Their cost, however, does not impose a penalty on control effort.

Chapter 6

Stochastic Adaptive Regulation

In this short chapter we address the stabilization problem for the system

$$dx = f(x)dt + g_1(x)dw + g_2(x)udt, \qquad (6.1)$$

where w is an independent Wiener process with incremental covariance $\Sigma\Sigma^{\mathrm{T}} dt$ and $g_1(0) = 0$. For the sake of discussion, let us assume that Σ is constant. For deterministic systems with constant parameters, the adaptive approach (see Chapter 6) allows the treatment of unknown parameters multiplying known nonlinearities. In the stochastic case we have the unknown noise dw multiplying the known nonlinearity $g_1(x)$. As we shall see in this chapter, the presence of noise does not prevent stabilization as long as $g_1(0) = 0$, i.e., as long as the equilibrium is preserved in the presence of noise. This is a strong condition which is usually not imposed in the so-called "stochastic (linear) adaptive control", where the noise is additive and non-vanishing (see, e.g. [22] and the reference therein). However, in the problem pursued here, the additional generality is that the noise can be of unknown (and, in fact, time-varying) covariance and it can multiply a nonlinearity.

6.1 Adaptive Stochastic Backstepping Design

In this chapter we deal with strict-feedback systems

$$dx_i = x_{i+1}dt + \varphi_i(\bar{x}_i)^{\mathrm{T}} dw, \qquad i = 1, \cdots, n-1 \qquad (6.2)$$
$$dx_n = udt + \varphi_n(\bar{x}_n)^{\mathrm{T}} dw, \qquad (6.3)$$

where w is an r-dimensional independent Wiener process with incremental covariance $\Sigma(t)\Sigma(t)^{\mathrm{T}} dt$ (as in Section 4.3), with an additional assumption that

$$\varphi_i(0) = 0. \qquad (6.4)$$

As we shall see in the sequel, to achieve adaptive stabilization in the presence of unknown Σ, for this class of systems, it is not necessary to estimate the entire matrix Σ and, in fact, it is possible to allow Σ to be time-varying. Instead we will estimate only one unknown parameter $\theta = \| \Sigma\Sigma^T \|_\infty$ using the estimate $\hat{\theta}(t)$. We employ the adaptive backstepping technique with tuning functions. Our presentation is very concise: instead of introducing the stabilizing functions $\alpha_i(\bar{x}_i, \hat{\theta})$ and tuning functions $\tau_i(\bar{x}_i, \hat{\theta})$ in a step-by-step fashion, we derive them simultaneously.

Since $\varphi_i(0) = 0$, the α_i's will vanish at $\bar{x}_i = 0$, as well as at $\bar{z}_i = 0$, where $\bar{z}_i = [z_1, \cdots, z_i]^T$. Thus, by the mean value theorem, $\alpha_i(\bar{x}_i, \hat{\theta})$ can be expressed as

$$\alpha_i(\bar{x}_i, \hat{\theta}) = \sum_{l=1}^{i} z_l \alpha_{il}(\bar{x}_i, \hat{\theta}). \tag{6.5}$$

where $\alpha_{il}(\bar{x}_i, \hat{\theta})$ are smooth functions. By introducing $z_i = x_i - \alpha_{i-1}$, we can now write $\varphi_i(\bar{x}_i)$ as

$$\varphi_i(\bar{x}_i) = \sum_{k=1}^{i} z_k \psi_{ik}(\bar{x}_i, \hat{\theta}), \tag{6.6}$$

where $\psi_{ik}(\bar{x}_i, \hat{\theta})$ are smooth functions. Then, according to Itô's differentiation rule, the system (6.2), (6.3) can be written as

$$\begin{aligned} dz_i &= d(x_i - \alpha_{i-1}) \\ &= \left(x_{i+1} - \sum_{l=1}^{i-1} \frac{\partial \alpha_{i-1}}{\partial x_l} x_{l+1} - \frac{1}{2} \sum_{p,q=1}^{i-1} \frac{\partial^2 \alpha_{i-1}}{\partial x_p \partial x_q} \varphi_p^T \Sigma\Sigma^T \varphi_q - \frac{\partial \alpha_{i-1}}{\partial \hat{\theta}} \dot{\hat{\theta}} \right) dt \\ &\quad + \left(\varphi_i^T - \sum_{l=1}^{i-1} \frac{\partial \alpha_{i-1}}{\partial x_l} \varphi_l^T \right) dw, \quad i = 1, \cdots, n, \end{aligned} \tag{6.7}$$

where $x_{n+1} = u$. We employ a Lyapunov function of the form

$$V(z, \tilde{\theta}) = \sum_{i=1}^{n} \frac{1}{4} z_i^4 + \frac{1}{2\gamma} \tilde{\theta}^2, \tag{6.8}$$

where $\tilde{\theta} = \| \Sigma\Sigma^T \|_\infty - \hat{\theta}$ is the parameter estimation error and $\hat{\theta}(t)$ is governed by the update law $\dot{\hat{\theta}} = \gamma \tau_n(x, \hat{\theta})$, and set out to select the functions $\alpha_i(\bar{x}_i, \hat{\theta})$ and $\tau_n(x, \hat{\theta})$ to make $\mathcal{L}V(z, \tilde{\theta})$ nonnegative. Along the solutions of (6.7), we have

$$\begin{aligned} \mathcal{L}V &= \sum_{i=1}^{n} z_i^3 \left(x_{i+1} - \sum_{l=1}^{i-1} \frac{\partial \alpha_{i-1}}{\partial x_l} x_{l+1} - \frac{1}{2} \sum_{p,q=1}^{i-1} \frac{\partial^2 \alpha_{i-1}}{\partial x_p \partial x_q} \varphi_p^T \Sigma\Sigma^T \varphi_q - \frac{\partial \alpha_{i-1}}{\partial \hat{\theta}} \dot{\hat{\theta}} \right) \\ &\quad + \frac{3}{2} \sum_{i=1}^{n} z_i^2 \left(\varphi_i - \sum_{l=1}^{i-1} \frac{\partial \alpha_{i-1}}{\partial x_l} \varphi_l \right)^T \Sigma\Sigma^T \left(\varphi_i - \sum_{l=1}^{i-1} \frac{\partial \alpha_{i-1}}{\partial x_l} \varphi_l \right) - \frac{\tilde{\theta} \dot{\hat{\theta}}}{\gamma} \end{aligned}$$

6.1 ADAPTIVE STOCHASTIC BACKSTEPPING DESIGN

$$\leq \sum_{i=1}^{n} z_i^3 \left(\alpha_i - \sum_{l=1}^{i-1} \frac{\partial \alpha_{i-1}}{\partial x_l} x_{l+1} - \frac{\partial \alpha_{i-1}}{\partial \hat{\theta}} \dot{\hat{\theta}} \right) + \sum_{i=1}^{n} z_i^3 z_{i+1}$$

$$- \frac{1}{2} \sum_{i=1}^{n} z_i^3 \sum_{p,q=1}^{i-1} \frac{\partial^2 \alpha_{i-1}}{\partial x_p \partial x_q} \varphi_p^T \Sigma \Sigma^T \varphi_q + \frac{3}{2} \sum_{i=1}^{n} z_i^2 \left(\varphi_i - \sum_{l=1}^{i-1} \frac{\partial \alpha_{i-1}}{\partial x_l} \varphi_l \right)^T$$

$$\times \left(\varphi_i - \sum_{l=1}^{i-1} \frac{\partial \alpha_{i-1}}{\partial x_l} \varphi_l \right) \| \Sigma \Sigma^T \|_\infty - \frac{\tilde{\theta} \dot{\hat{\theta}}}{\gamma}. \tag{6.9}$$

Consider the third term,

$$-\frac{1}{2} \sum_{i=1}^{n} z_i^3 \sum_{p,q=1}^{i-1} \frac{\partial^2 \alpha_{i-1}}{\partial x_p \partial x_q} \varphi_p^T \Sigma \Sigma^T \varphi_q$$

$$\leq \frac{1}{2} \sum_{i=1}^{n} \sum_{p,q=1}^{i-1} |z_i|^3 \left| \frac{\partial^2 \alpha_{i-1}}{\partial x_p \partial x_q} \right| \left| \sum_{k=1}^{p} z_k \psi_{pk} \right| \left| \sum_{l=1}^{q} z_l \psi_{ql} \right| \left| \Sigma \Sigma^T \right|$$

$$\leq \frac{1}{2} \sum_{i=1}^{n} \sum_{p,q=1}^{i-1} \sum_{k=1}^{p} \sum_{l=1}^{q} \left| \frac{\partial^2 \alpha_{i-1}}{\partial x_p \partial x_q} \right| |\psi_{pk}| |\psi_{ql}| |z_i|^3 |z_k| |z_l| \left| \Sigma \Sigma^T \right|$$

$$\leq \frac{1}{4} \sum_{i=1}^{n} \sum_{p,q=1}^{i-1} \sum_{k=1}^{p} \sum_{l=1}^{q} \left\{ \left(\frac{\partial^2 \alpha_{i-1}}{\partial x_p \partial x_q} \right)^2 |\psi_{pk}|^2 |\psi_{ql}|^2 z_i^6 + \frac{1}{2} z_k^4 + \frac{1}{2} z_l^4 \right\} \left| \Sigma \Sigma^T \right|$$

$$= \frac{1}{4} \sum_{i=1}^{n} z_i^6 \sum_{p,q=1}^{i-1} \sum_{k=1}^{p} \sum_{l=1}^{q} \left(\frac{\partial^2 \alpha_{i-1}}{\partial x_p \partial x_q} \right)^2 \psi_{pk}^T \psi_{pk} \psi_{ql}^T \psi_{ql} \left| \Sigma \Sigma^T \right|$$

$$+ \frac{1}{16} \sum_{i=1}^{n-1} z_i^4 \sum_{k=i+1}^{n} k(k-1)(k-i)(k+i-1) \left| \Sigma \Sigma^T \right|, \tag{6.10}$$

and employing the inequalities (3.42) and (3.44)–(3.49), we have

$$\mathcal{L}V \leq \sum_{i=1}^{n} z_i^3 \left(\alpha_i - \sum_{l=1}^{i-1} \frac{\partial \alpha_{i-1}}{\partial x_l} x_{l+1} - \frac{\partial \alpha_{i-1}}{\partial \hat{\theta}} \dot{\hat{\theta}} \right) + \frac{3}{4} \sum_{i=1}^{n} \epsilon_i^{\frac{4}{3}} z_i^4 + \sum_{i=1}^{n} \frac{1}{4 \epsilon_{i-1}^4} z_i^4$$

$$+ \frac{1}{4} \sum_{i=1}^{n} z_i^6 \sum_{p,q=1}^{i-1} \sum_{k=1}^{p} \sum_{l=1}^{q} \left(\frac{\partial^2 \alpha_{i-1}}{\partial x_p \partial x_q} \right)^2 \psi_{pk}^T \psi_{pk} \psi_{ql}^T \psi_{ql} \| \Sigma \Sigma^T \|_\infty$$

$$+ \frac{1}{16} \sum_{i=1}^{n-1} z_i^4 \sum_{k=i+1}^{n} k(k-1)(k-i)(k+i-1) \| \Sigma \Sigma^T \|_\infty$$

$$+ \left[\frac{3}{2} \sum_{i=1}^{n} \left(z_i^4 \beta_{ii}^T \beta_{ii} + 2 z_i^3 \beta_{ii}^T \sum_{k=1}^{i-1} z_k \beta_{ik} \right) + \frac{3}{4} \sum_{i=1}^{n} z_i^4 \left(\sum_{j=1}^{r} \sum_{k=1}^{i-1} \sum_{l=1}^{i-1} \frac{1}{\epsilon_{ikl}^2} \beta_{ikj}^2 \beta_{ilj}^2 \right) \right.$$

$$\left. + \frac{3r}{4} \sum_{i=1}^{n-1} z_i^4 \left(\sum_{k=i+1}^{n} \sum_{l=1}^{k-1} \epsilon_{kil}^2 \right) \right] (\hat{\theta} + \tilde{\theta}) - \frac{\tilde{\theta} \dot{\hat{\theta}}}{\gamma}$$

$$= \sum_{i=1}^{n} z_i^3 \left[\alpha_i - \sum_{l=1}^{i-1} \frac{\partial \alpha_{i-1}}{\partial x_l} x_{l+1} + \frac{1}{4} z_i^3 \sum_{p,q=1}^{i-1} \sum_{k=1}^{p} \sum_{l=1}^{q} \left(\frac{\partial^2 \alpha_{i-1}}{\partial x_p \partial x_q} \right)^2 \psi_{pk}^T \psi_{pk} \psi_{ql}^T \psi_{ql} \hat{\theta} \right.$$

$$+\frac{1}{16}z_i\sum_{k=i+1}^{n}k(k-1)(k-i)(k+i-1)\hat{\theta}+\frac{3}{4}\epsilon_i^{\frac{4}{3}}z_i+\frac{1}{4\epsilon_{i-1}^4}z_i+\left(\frac{3}{2}z_i\beta_{ii}^{\mathrm{T}}\beta_{ii}\right.$$

$$+3\beta_{ii}^{\mathrm{T}}\sum_{k=1}^{i-1}z_k\beta_{ik}+\frac{3}{4}z_i\sum_{j=1}^{r}\sum_{k=1}^{i-1}\sum_{l=1}^{i-1}\frac{1}{\epsilon_{ikl}^2}\beta_{ikj}^2\beta_{ilj}^2+\frac{3r}{4}z_i\sum_{k=i+1}^{n}\sum_{l=1}^{k-1}\epsilon_{kil}^2\left.\right)\hat{\theta}$$

$$-\frac{\partial\alpha_{i-1}}{\partial\hat{\theta}}\dot{\hat{\theta}}\Bigg]-\tilde{\theta}\left[\frac{\dot{\hat{\theta}}}{\gamma}-\frac{1}{4}\sum_{i=1}^{n}z_i^6\sum_{p,q=1}^{i-1}\sum_{k=1}^{p}\sum_{l=1}^{q}\left(\frac{\partial^2\alpha_{i-1}}{\partial x_p\partial x_q}\right)^2\psi_{pk}^{\mathrm{T}}\psi_{pk}\psi_{ql}^{\mathrm{T}}\psi_{ql}\right.$$

$$-\frac{1}{16}\sum_{i=1}^{n}z_i^4\sum_{k=i+1}^{n}k(k-1)(k-i)(k+i-1)$$

$$-\frac{3}{2}\sum_{i=1}^{n}\left(z_i^4\beta_{ii}^{\mathrm{T}}\beta_{ii}+2z_i^3\beta_{ii}^{\mathrm{T}}\sum_{k=1}^{i-1}z_k\beta_{ik}\right)-\frac{3}{4}\sum_{i=1}^{n}z_i^4\left(\sum_{j=1}^{r}\sum_{k=1}^{i-1}\sum_{l=1}^{i-1}\frac{1}{\epsilon_{ikl}^2}\beta_{ikj}^2\beta_{ilj}^2\right)$$

$$-\frac{3r}{4}\sum_{i=1}^{n-1}z_i^4\left(\sum_{k=i+1}^{n}\sum_{l=1}^{k-1}\epsilon_{kil}^2\right)\Bigg], \tag{6.11}$$

where $x_{n+1}=u$, $z_{n+1}=0$, $\alpha_n=u$, $\epsilon_n=0$ and $\epsilon_0=\infty$. Let

$$\dot{\hat{\theta}} = \gamma\tau_n \tag{6.12}$$
$$\tau_i = \tau_{i-1}+\omega_iz_i^3, \qquad i=1,\cdots,n, \tag{6.13}$$

where $\tau_0=0$ and

$$\begin{aligned}\omega_i =\ & \frac{1}{4}z_i^3\sum_{p,q=1}^{i-1}\sum_{k=1}^{p}\sum_{l=1}^{q}\left(\frac{\partial^2\alpha_{i-1}}{\partial x_p\partial x_q}\right)^2\psi_{pk}^{\mathrm{T}}\psi_{pk}\psi_{ql}^{\mathrm{T}}\psi_{ql}\\ &+\frac{1}{16}z_i\sum_{k=i+1}^{n}k(k-1)(k-i)(k+i-1)\\ &+\frac{3}{2}z_i\beta_{ii}^{\mathrm{T}}\beta_{ii}+3\beta_{ii}^{\mathrm{T}}\sum_{k=1}^{i-1}z_k\beta_{ik}+\frac{3}{4}z_i\left(\sum_{j=1}^{r}\sum_{k=1}^{i-1}\sum_{l=1}^{i-1}\frac{1}{\epsilon_{ikl}^2}\beta_{ikj}^2\beta_{ilj}^2\right)\\ &+\frac{3r}{4}z_i\sum_{k=i+1}^{n}\sum_{l=1}^{k-1}\epsilon_{kil}^2.\end{aligned}\tag{6.14}$$

Then

$$\begin{aligned}\mathcal{L}V \le\ & \sum_{i=1}^{n}z_i^3\left(\alpha_i-\sum_{l=1}^{i-1}\frac{\partial\alpha_{i-1}}{\partial x_l}x_{l+1}+\frac{3}{4}\epsilon_i^{\frac{4}{3}}z_i+\frac{1}{4\epsilon_{i-1}^4}z_i+\omega_i\hat{\theta}-\frac{\partial\alpha_{i-1}}{\partial\hat{\theta}}\sum_{j=1}^{n}\gamma z_j^3\omega_j\right)\\ =\ & \sum_{i=1}^{n}z_i^3\left(\alpha_i-\sum_{l=1}^{i-1}\frac{\partial\alpha_{i-1}}{\partial x_l}x_{l+1}+\frac{3}{4}\epsilon_i^{\frac{4}{3}}z_i+\frac{1}{4\epsilon_{i-1}^4}z_i+\omega_i\hat{\theta}-\frac{\partial\alpha_{i-1}}{\partial\hat{\theta}}\sum_{j=1}^{i}\gamma z_j^3\omega_j\right)\\ & -\sum_{i=1}^{n-1}z_i^3\frac{\partial\alpha_{i-1}}{\partial\hat{\theta}}\sum_{j=i+1}^{n}\gamma z_j^3\omega_j\end{aligned}$$

6.1 ADAPTIVE STOCHASTIC BACKSTEPPING DESIGN

$$= \sum_{i=1}^{n} z_i^3 \left(\alpha_i - \sum_{l=1}^{i-1} \frac{\partial \alpha_{i-1}}{\partial x_l} x_{l+1} + \frac{3}{4} \epsilon_i^{\frac{4}{3}} z_i + \frac{1}{4\epsilon_{i-1}^4} z_i + \omega_i \hat{\theta} - \frac{\partial \alpha_{i-1}}{\partial \hat{\theta}} \sum_{j=1}^{i} \gamma z_j^3 \omega_j \right)$$

$$- \sum_{i=2}^{n} z_i^3 \sum_{j=1}^{i-1} \frac{\partial \alpha_{j-1}}{\partial \hat{\theta}} \gamma z_j^3 \omega_i$$

$$= \sum_{i=1}^{n} z_i^3 \left(\alpha_i - \sum_{l=1}^{i-1} \frac{\partial \alpha_{i-1}}{\partial x_l} x_{l+1} + \frac{3}{4} \epsilon_i^{\frac{4}{3}} z_i + \frac{1}{4\epsilon_{i-1}^4} z_i + \omega_i \hat{\theta} - \frac{\partial \alpha_{i-1}}{\partial \hat{\theta}} \sum_{j=1}^{i} \gamma z_j^3 \omega_j \right.$$

$$\left. - \sum_{j=1}^{i-1} \frac{\partial \alpha_{j-1}}{\partial \hat{\theta}} \gamma z_j^3 \omega_i \right). \tag{6.15}$$

Letting

$$\alpha_i = -c_i z_i + \sum_{l=1}^{i-1} \frac{\partial \alpha_{i-1}}{\partial x_l} x_{l+1} - \frac{3}{4} \epsilon_i^{\frac{4}{3}} z_i - \frac{1}{4\epsilon_{i-1}^4} z_i - \omega_i \hat{\theta} + \frac{\partial \alpha_{i-1}}{\partial \hat{\theta}} \sum_{j=1}^{i} \gamma z_j^3 \omega_j$$

$$+ \sum_{j=1}^{i-1} \frac{\partial \alpha_{j-1}}{\partial \hat{\theta}} \gamma z_j^3 \omega_i \tag{6.16}$$

$$u = \alpha_n, \tag{6.17}$$

where $c_i > 0$, the infinitesimal generator of the system (6.7) becomes negative definite:

$$\mathcal{L}V(z, \tilde{\theta}) \leq - \sum_{i=1}^{n} c_i z_i^4. \tag{6.18}$$

Since $z = 0 \Rightarrow x = 0$, by Theorem 3.3 we have the following result.

Theorem 6.1 *The equilibrium $x = 0$, $\hat{\theta} = \| \Sigma \Sigma^T \|_\infty$ of the closed-loop system (6.2), (6.3), (6.12) and (6.17) is globally stable in probability and*

$$P\left\{ \lim_{t \to \infty} x(t) = 0 \right\} = 1. \tag{6.19}$$

Remark 6.2 Since $\mathcal{L}V$ is nonpositive, $EV(t)$ is nonincreasing. Since V is also bounded from below by zero, $EV(t)$ has a limit. Since $x(t)$, and therefore $z(t)$, converges to zero with probability one, $E\{z(t)^4\} \to 0$, and thus $E\{\tilde{\theta}(t)^2\}$ has a limit. Denote this limit by $\Delta \theta^2$. Thus

$$\text{l.i.m.}_{t \to \infty} \hat{\theta}(t) = \| \Sigma \Sigma^T \|_\infty \pm \Delta \theta. \tag{6.20}$$

It is well known [18] that *convergence in the mean* implies *convergence in probability*, i.e.

$$\lim_{t \to \infty} P\left\{ \left| \hat{\theta}(t) - \left(\| \Sigma \Sigma^T \|_\infty + \Delta \theta \right) \right| > \epsilon \right\} P\left\{ \left| \hat{\theta}(t) - \left(\| \Sigma \Sigma^T \|_\infty - \Delta \theta \right) \right| > \epsilon \right\} = 0,$$

$$\tag{6.21}$$

$\forall \epsilon > 0$. □

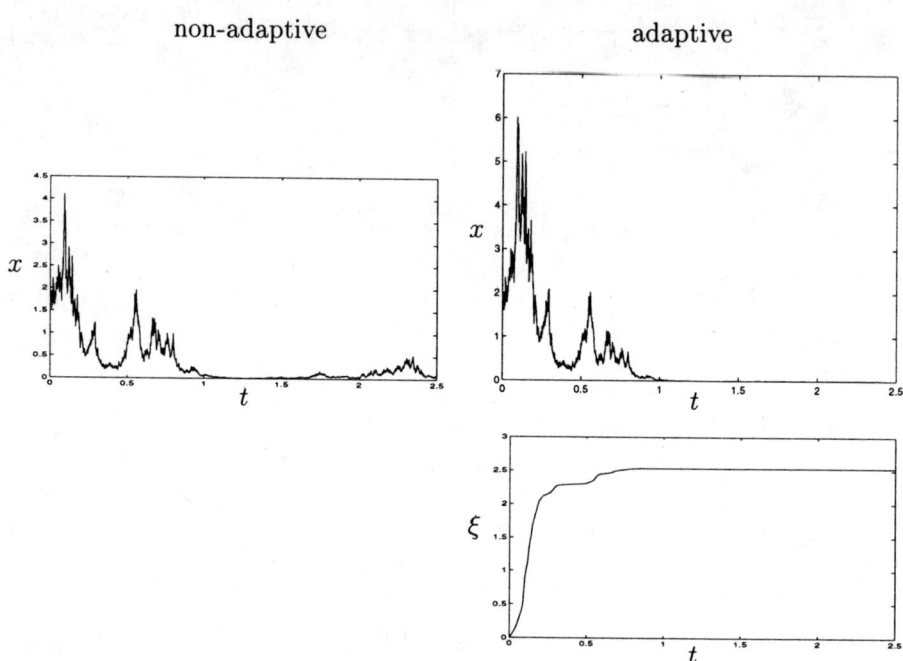

Figure 6.1: The time responses with the non-adaptive and the adaptive controller.

6.2 Example

This example gives an idea about the difference between the noise-to-state stabilization from Chapter 4 and the adaptive stabilization from this chapter. Consider the scalar system

$$dx = udt + xdw, \tag{6.22}$$

where w is an independent Wiener process with $E\{dw^2\} = \sigma(t)^2 dt$. Consider the following two control laws:

$$u = -x - x^3 \tag{6.23}$$
$$u = -x - \xi x, \quad \dot{\xi} = x^2. \tag{6.24}$$

It can be shown that they guarantee, respectively, that

$$E\{x(t)^2\} \leq e^{-2t} E\{x(0)^2\} + \frac{1}{16} \sup_{0 \leq \tau \leq t} \sigma(\tau)^4 \tag{6.25}$$

$$\frac{d}{dt} E\left\{ x^2 + \left(\frac{\|\sigma^2\|_\infty}{2} - \xi \right)^2 \right\} \leq -2E\{x^2\}. \tag{6.26}$$

The controller (6.23) is a disturbance attenuation controller which guarantees NSS. The controller (6.24) is an adaptive controller which guarantees that $x = \xi = 0$ is globally stable and x is regulated to zero (in probability). The $x(t)$ time responses in Figure 6.1 reveal the difference between the achieved stability properties. The simulations are performed for $\sigma(t) \equiv 2\sqrt{2}$. While the adaptive controller on the right achieves regulation of x, the non-adaptive controller on the left achieves only noise-to-state stability, which means that it forces x to converge to an interval around zero proportional to σ. As evident from the figure, the non-adaptive controller results in a *residual error*, whereas the adaptive controller does not. The variable ξ is the estimate of $\|\sigma^2\|_\infty/2 = 4$. We see that $\xi(t)$ converges to about 2.5 and does not reach the true value 4. This is not unexpected as in adaptive regulation problems we seldom see convergence to the true parameter.

Notes and References

When applied to linear systems, the design in this chapter solves the stabilization problem with multiplicative noise. A sizeable body of literature on this problem was reviewed by Bernstein [8]. All of the previous results assume either restrictive geometric conditions as, e.g., in Willems and Willems [113] or require the knowledge of a bound on the noise covariance as in Wonham [23] and El Ghaoui [23]. The adaptive design requires no a priori knowledge of a bound on the covariance (it "tunes" itself to the bound) and applies to a class of systems that does not satisfy the geometric conditions in [113].

Chapter 7

Invariant Manifolds and Asymptotic Properties

With its focus on *optimality*, this book devotes considerable effort to *transient* performance of control systems. This chapter addresses *asymptotic* performance of adaptive systems.

While in the case of persistent excitation (PE) the parameter estimates in adaptive control converge to the actual parameter values, very little is known about asymptotic behavior of adaptive controllers in the absence of PE. A classical question is: does the parameter estimate $\hat{\theta}(t)$ converge to a *constant* value when there is no PE? Even though in every adaptive system $\dot{\hat{\theta}}(t) \to 0$ as $t \to \infty$, this does not guarantee that $\hat{\theta}(t)$ has a limit, as, e.g., in the case of the function $\hat{\theta}(t) = \sin(\ln(1+t))$. The only estimation algorithm which is known to guarantee convergence of $\hat{\theta}(t)$ to a constant in the absence of PE is the least-squares algorithm. There are no convergence guarantees for the gradient and the Lyapunov-type algorithms.

In this chapter we address the problem of convergence of the parameter estimates for the case of set point regulation in the tuning functions design from Section 5.6. While the extension to other adaptive schemes does not seem to be difficult, the extension to tracking arbitrary trajectories appears plagued by insurmountable obstacles.

The approach in the analysis is the invariant manifold theory which allows to draw conclusions about the sizes of sets leading to stabilizing or destabilizing parameter values. The results of the chapter are quite surprising. If we consider the rank of the regressor matrix at the equilibrium of the system to be an indicator of the level of excitation in the system, then the findings are as follows:

1. *persistent excitation* always guarantees convergence to stabilizing parameters;

2. *least excitation* (where the regressor is zero at the equilibrium) allows

convergence to destabilizing values from a set of initial conditions of Lebesgue measure zero;

3. *partial excitation* (where the regressor is neither zero nor with a full rank at the equilibrium) allows convergence to destabilizing values from a set of initial conditions of positive Lebesgue measure.

Thus the dependence of the convergence properties on the excitation level is not monotonic. Moreover, these results disprove the intuition that adaptive stabilization mechanism consists of the convergence of the parameter estimate to a stabilizing value.

7.1 Closed-Loop Adaptive System

We briefly review the adaptive backstepping design with tuning functions to arrive at its closed-loop system. Consider nonlinear systems in the *parametric strict-feedback form*:

$$\begin{aligned}
\dot{x}_1 &= x_2 + \varphi_1(x_1)^\mathrm{T}\theta \\
\dot{x}_2 &= x_3 + \varphi_2(x_1, x_2)^\mathrm{T}\theta \\
&\vdots \\
\dot{x}_{n-1} &= x_n + \varphi_{n-1}(x_1, \ldots, x_{n-1})^\mathrm{T}\theta \\
\dot{x}_n &= u + \varphi_n(x)^\mathrm{T}\theta
\end{aligned} \quad (7.1)$$

where $\theta \in \mathbb{R}^p$ is the vector of unknown constant parameters and the elements of $F = [\varphi_1, \cdots, \varphi_n]$ are smooth nonlinear functions taking arguments in $\mathbb{R}^n$. We consider the problem of adaptive regulation of the output $y = x_1$ to a given set-point y_s. Starting with $x_1^e = y_s$, we solve the n equilibrium equations of (7.1) to get

$$\begin{aligned}
x_1^e &= y_s \\
x_2^e &= -\varphi_1(x_1^e)^\mathrm{T}\theta \\
x_3^e &= -\varphi_2(x_1^e, x_2^e)^\mathrm{T}\theta \\
&\vdots \\
x_n^e &= -\varphi_{n-1}(x_1^e, \ldots, x_{n-1}^e)^\mathrm{T}\theta.
\end{aligned} \quad (7.2)$$

The control objective is to globally stabilize the equilibrium $x^e = [x_1^e, \cdots, x_n^e]^\mathrm{T}$ for any unknown value of θ.

7.1 CLOSED-LOOP ADAPTIVE SYSTEM

For conceptual simplicity, we now take the simplest control law that achieves the stabilization objective. We take the tuning functions control law from [63, Section 4.2]. The control law in Section 5.6 is more complicated because it avoids cancellation of nonlinearities. The control procedure in [63, Section 4.2] is given recursively using the expressions

$$z_i = x_i - \alpha_{i-1} \tag{7.3}$$

$$\alpha_i(\bar{x}_i, \hat{\theta}) = -z_{i-1} - c_i z_i - w_i^T \hat{\theta} + \sum_{k=1}^{i-1} \frac{\partial \alpha_{i-1}}{\partial x_k} x_{k+1}$$

$$+ \frac{\partial \alpha_{i-1}}{\partial \hat{\theta}} \Gamma \tau_i + \sum_{k=2}^{i-1} \frac{\partial \alpha_{k-1}}{\partial \hat{\theta}} \Gamma w_i z_k \tag{7.4}$$

$$\tau_i(\bar{x}_i, \hat{\theta}) = \tau_{i-1} + w_i z_i \tag{7.5}$$

$$w_i(\bar{x}_i, \hat{\theta}) = \varphi_i - \sum_{k=1}^{i-1} \frac{\partial \alpha_{i-1}}{\partial x_k} \varphi_k \tag{7.6}$$

$$\bar{x}_i = (x_1, \ldots, x_i) \qquad i = 1, \ldots, n,$$

where $\alpha_0 = 0$, $\tau_0 = 0$, and $\Gamma = \Gamma^T > 0$. The control law is

$$u = \alpha_n(x, \tilde{\theta}), \tag{7.7}$$

and the adaptive law is

$$\dot{\hat{\theta}} = \Gamma \tau_n(x, \hat{\theta}) = \Gamma W(z, \tilde{\theta}) z, \tag{7.8}$$

where $W(z, \tilde{\theta}) = [w_1, \cdots, w_n] = F(x) N(z, \tilde{\theta})^T$ and

$$N(z, \tilde{\theta}) = \begin{bmatrix} 1 & 0 & \cdots & 0 \\ -\frac{\partial \alpha_1}{\partial x_1} & 1 & \ddots & \vdots \\ \vdots & \ddots & \ddots & 0 \\ -\frac{\partial \alpha_{n-1}}{\partial x_1} & \cdots & -\frac{\partial \alpha_{n-1}}{\partial x_{n-1}} & 1 \end{bmatrix}. \tag{7.9}$$

This adaptive controller results in a closed-loop system of the form

$$\dot{z} = A_z(z, \tilde{\theta}) z + W(z, \tilde{\theta})^T \tilde{\theta} \tag{7.10}$$

$$\dot{\tilde{\theta}} = -\Gamma W(z, \tilde{\theta}) z \tag{7.11}$$

where $\tilde{\theta} = \theta - \hat{\theta}$ is the parameter estimation error, and

$$A_z(z, \tilde{\theta}) = \begin{bmatrix} -c_1 & 1 & 0 & \cdots & 0 \\ -1 & -c_2 & 1+\sigma_{23} & \cdots & \sigma_{2n} \\ 0 & -1-\sigma_{23} & \ddots & \ddots & \vdots \\ \vdots & \vdots & \ddots & \ddots & 1+\sigma_{n-1,n} \\ 0 & -\sigma_{2n} & \cdots & -1-\sigma_{n-1,n} & -c_n \end{bmatrix} \tag{7.12}$$

and
$$\sigma_{jk}(z,\tilde{\theta}) = -\frac{\partial \alpha_{j-1}}{\partial \hat{\theta}}\Gamma w_k. \tag{7.13}$$

Let us denote $F_e = F(x^e)$ and $r = \text{rank}\{F_e\}$. Then we have the following theorem.

Theorem 7.1 (Theorem 4.12 in [63]) *The closed-loop adaptive system (7.10), (7.11), has a globally stable equilibrium $(z,\tilde{\theta}) = 0$. Furthermore, its state $\left(z(t),\tilde{\theta}(t)\right)$ converges to the $(p-r)$-dimensional equilibrium manifold M given by*

$$M = \left\{(z,\tilde{\theta}) \in \mathbb{R}^{n+p} | z = 0, F_e^T \tilde{\theta} = 0\right\}. \tag{7.14}$$

An important property of M is its dimension, $p - r$:

1. When $r = p$, i.e., $\dim\{M\} = 0$, M becomes the equilibrium point $z = 0$, $\tilde{\theta} = 0$. This equilibrium is globally asymptotically stable and the parameter estimate $\hat{\theta}(t)$ converges to its true value θ. This is the case of *persistent excitation*.

2. When $F_e = 0$, i.e., $\dim\{M\} = p$, M becomes the equilibrium manifold $z = 0$. This is the case of *least excitation*.

3. When $0 < r < p$, we have *partial excitation*. This is the general case which is harder to analyze than the two limiting cases.

7.2 Parameter Estimates Converge to Constant Values

The first difficulty in studying asymptotic properties of adaptive controllers is to prove that the parameter estimates converge to constant values.

Theorem 7.2 *Consider the adaptive system (7.10)–(7.11). There exists a constant vector $\hat{\theta}_\infty \in \mathbb{R}^p$ such that $\lim_{t\to\infty} \hat{\theta}(t) = \hat{\theta}_\infty$.*

Proof. Consider the Lyapunov function

$$V_n = \frac{1}{2}z^T z + \frac{1}{2}\tilde{\theta}^T \Gamma^{-1}\tilde{\theta}. \tag{7.15}$$

In view of (7.12), the derivative of V_n along (7.10)–(7.11) is

$$\dot{V}_n = -\sum_{k=1}^{n} c_k z_k^2 \leq -c_0 |z|^2, \tag{7.16}$$

7.2 PARAMETER ESTIMATES CONVERGE TO CONSTANT VALUES

where $c_0 = \min\{c_1, \ldots, c_n\}$. Then it follows that $z \in \mathcal{L}_2$. First, we show that $F_e^T \tilde{\theta} \in \mathcal{L}_2$. To do this, we use induction.

Consider the z_1-equation:

$$\begin{aligned} \dot{z}_1 &= -c_1 z_1 + z_2 + \varphi_1(x_1)^T \tilde{\theta} \\ &= -c_1 z_1 + z_2 + \varphi_1(x_1^e)^T \tilde{\theta} + (\varphi_1(x_1) - \varphi_1(x_1^e))^T \tilde{\theta}. \end{aligned} \quad (7.17)$$

Since $z_1 = x_1 - x_1^e \in \mathcal{L}_2$, we have that $\varphi_1(x_1) - \varphi_1(x_1^e) \in \mathcal{L}_2$. In the following text, we use $[\mathcal{L}_2]$ as a generic expression for $\mathcal{L}_2$ terms. Since $z_2 \in \mathcal{L}_2$, (7.17) becomes

$$\dot{z}_1 = -c_1 z_1 + \varphi_1(x_1^e)^T \tilde{\theta} + [\mathcal{L}_2]. \quad (7.18)$$

On the other hand, we have

$$\begin{aligned} \varphi_1(x_1^e)^T \dot{\tilde{\theta}} &= -\varphi_1(x_1^e)^T \Gamma F(x) N^T z \\ &= -\varphi_1(x_1^e)^T \Gamma \varphi_1(x_1^e) z_1 - \varphi_1(x_1^e)^T \Gamma \left(F N^T - \varphi_1(x_1^e) e_1^T \right) z \\ &= -\varphi_1(x_1^e)^T \Gamma \varphi_1(x_1^e) z_1 + [\mathcal{L}_2] \end{aligned} \quad (7.19)$$

due to $z \in \mathcal{L}_2$. Combining (7.18) with (7.19) we get

$$\begin{bmatrix} \dot{z}_1 \\ \varphi_1(x_1^e)^T \dot{\tilde{\theta}} \end{bmatrix} = \begin{bmatrix} -c_1 & 1 \\ -\varphi_1(x_1^e)^T \Gamma \varphi_1(x_1^e) & 0 \end{bmatrix} \begin{bmatrix} z_1 \\ \varphi_1(x_1^e)^T \tilde{\theta} \end{bmatrix} + [\mathcal{L}_2]. \quad (7.20)$$

If $\varphi_1(x_1^e) \neq 0$, system (7.20) is a stable LTI system driven by an $\mathcal{L}_2$ signal, so that

$$\varphi_1(x_1^e)^T \tilde{\theta} \in \mathcal{L}_2 \quad (7.21)$$

(otherwise $\varphi_1(x_1^e)^T \tilde{\theta} = 0 \in \mathcal{L}_2$). This, in turn, means that $\varphi_1(x_1)^T \tilde{\theta} \in \mathcal{L}_2$, $\dot{z}_1 \in \mathcal{L}_2$ and $\dot{x}_1 \in \mathcal{L}_2$.

We then assume that at step $i - 1$ we have proved that $x_k - x_k^e \in \mathcal{L}_2$, $\varphi_k(\bar{x}_k) - \varphi_k(\bar{x}_k^e) \in \mathcal{L}_2$, $\varphi_k(\bar{x}_k^e)^T \tilde{\theta} \in \mathcal{L}_2$, $\varphi_k(\bar{x}_k)^T \tilde{\theta} \in \mathcal{L}_2$, $\dot{z}_k \in \mathcal{L}_2$, $\dot{x}_k \in \mathcal{L}_2$, $k = 1, \cdots, i - 1$. Now consider the z_i-equation:

$$\begin{aligned} \dot{z}_i &= -c_i z_i - z_{i-1} + z_{i+1} + \sum_{k=i+1}^{n} \sigma_{ik} z_k - \sum_{k=2}^{i-1} \sigma_{ki} z_k + w_i^T \tilde{\theta} \\ &= -c_i z_i + \varphi_i(\bar{x}_i)^T \tilde{\theta} - \sum_{k=1}^{i-1} \frac{\partial \alpha_{i-1}}{\partial x_k} \underbrace{\varphi_k(\bar{x}_k)^T \tilde{\theta}}_{\in \mathcal{L}_2 (\text{step } i-1)} + [\mathcal{L}_2] \\ &= -c_i z_i + \varphi_i(\bar{x}_i^e)^T \tilde{\theta} + (\varphi_i(\bar{x}_i) - \varphi_i(\bar{x}_i^e))^T \tilde{\theta} + [\mathcal{L}_2], \end{aligned} \quad (7.22)$$

in which we have used $z \in \mathcal{L}_2$ and $w_i = \varphi_i(\bar{x}_i) - \sum_{k=1}^{i-1} \frac{\partial \alpha_{i-1}}{\partial x_k} \varphi_k(\bar{x}_k)$. Since

$$x_i = \underbrace{\dot{x}_{i-1}}_{\in \mathcal{L}_2} - \varphi_{i-1}(\bar{x}_{i-1})^T \theta$$

$$= \underbrace{-\varphi_{i-1}(\bar{x}_{i-1}^e)^T\theta}_{x_i^e} - \underbrace{\left(\varphi_{i-1}(\bar{x}_{i-1}) - \varphi_{i-1}(\bar{x}_{i-1}^e)\right)^T\theta}_{\in \mathcal{L}_2(\text{step } i-1)} + [\mathcal{L}_2]$$

$$= x_i^e + [\mathcal{L}_2], \tag{7.23}$$

combining with the assumption $x_k - x_k^e \in \mathcal{L}_2, k = 1, \cdots, i-1$, it follows that

$$\varphi_i(\bar{x}_i) - \varphi_i(\bar{x}_i^e) = \varphi_i(\bar{x}_i^e + [\mathcal{L}_2]) - \varphi_i(\bar{x}_i^e) \in \mathcal{L}_2. \tag{7.24}$$

Therefore we have

$$\dot{z}_i = -c_i z_i + \varphi_i(\bar{x}_i^e)^T \tilde{\theta} + [\mathcal{L}_2]. \tag{7.25}$$

On the other hand,

$$\varphi_i(\bar{x}_i^e)^T \dot{\tilde{\theta}} = -\varphi_i(\bar{x}_i^e)^T \Gamma F(x) N^T z$$

$$= -\varphi_i(\bar{x}_i^e)^T \Gamma \varphi_i(\bar{x}_i^e) z_i - \varphi_i(\bar{x}_i^e)^T \Gamma \left(FN^T - \varphi_i(\bar{x}_i^e)e_i^T\right) z$$

$$= -\varphi_i(\bar{x}_i^e)^T \Gamma \varphi_i(\bar{x}_i^e) z_i + [\mathcal{L}_2]. \tag{7.26}$$

Combining (7.25) with (7.26), we get

$$\begin{bmatrix} \dot{z}_i \\ \varphi_i(\bar{x}_i^e)^T \dot{\tilde{\theta}} \end{bmatrix} = \begin{bmatrix} -c_i & 1 \\ -\varphi_i(\bar{x}_i^e)^T \Gamma \varphi_i(\bar{x}_i^e) & 0 \end{bmatrix} \begin{bmatrix} z_i \\ \varphi_i(\bar{x}_i^e)^T \tilde{\theta} \end{bmatrix} + [\mathcal{L}_2]. \tag{7.27}$$

If $\varphi_i(\bar{x}_i^e) \neq 0$, system (7.27) is a stable LTI system driven by an $\mathcal{L}_2$ signal, so that

$$\varphi_i(\bar{x}_i^e)^T \tilde{\theta} \in \mathcal{L}_2 \tag{7.28}$$

(otherwise $\varphi_i(\bar{x}_i^e)^T \tilde{\theta} = 0 \in \mathcal{L}_2$). This, in turn, means that $\varphi_i(\bar{x}_i)^T \tilde{\theta} \in \mathcal{L}_2$, $\dot{z}_i \in \mathcal{L}_2$. Also, note that

$$\dot{x}_i = \underbrace{\dot{z}_i}_{\in \mathcal{L}_2} - \sum_{k=1}^{i-1} \frac{\partial \alpha_{i-1}}{\partial z_k} \underbrace{\dot{z}_k}_{\in \mathcal{L}_2} - \frac{\partial \alpha_{i-1}}{\partial \hat{\theta}} \underbrace{\dot{\hat{\theta}}}_{\in \mathcal{L}_2} \in \mathcal{L}_2. \tag{7.29}$$

We conclude from the above induction that $x_i - x_i^e \in \mathcal{L}_2$, $\varphi_i(\bar{x}_i) - \varphi_i(\bar{x}_i^e) \in \mathcal{L}_2$, $\varphi_i(\bar{x}_i^e)^T \tilde{\theta} \in \mathcal{L}_2$, $\varphi_i(\bar{x}_i)^T \tilde{\theta} \in \mathcal{L}_2$, $\dot{z}_i \in \mathcal{L}_2$, $\dot{x}_i \in \mathcal{L}_2$ for $i = 1, \cdots, n$. Thus

$$\begin{bmatrix} \varphi_1(x_1^e)^T \\ \varphi_2(\bar{x}_2^e)^T \\ \vdots \\ \varphi_n(\bar{x}_n^e) \end{bmatrix} \tilde{\theta} = F_e^T \tilde{\theta} \in \mathcal{L}_2. \tag{7.30}$$

Now we finish the proof of the theorem. Let $r = \text{rank}\{F_e^T\}$ and define

$$\bar{P} = p \times r \text{ matrix of basis vectors of Range}\{F_e\},$$
$$\bar{Q} = r \times p \text{ matrix of basis vectors of Null}\{(\Gamma F_e)^T\}.$$

7.2 Parameter Estimates Converge to Constant Values

From Theorem 2.1, we have that $F_e^T \tilde{\theta}(t) \to 0$, so $\bar{P}^T \tilde{\theta}(t) \to 0$, i.e., $\bar{P}^T \hat{\theta}(t) \to$ *const*. On the other hand, noting (7.8) we have

$$\bar{Q}^T \dot{\hat{\theta}} = \bar{Q}^T \Gamma F N^T z$$
$$= \bar{Q}^T \left(\Gamma F_e N^T z + \Gamma(F - F_e) N^T z \right). \tag{7.31}$$

Since $x - x^e \in \mathcal{L}_2$, we note that $F(x) - F(x^e) \in \mathcal{L}_2$. Recalling that $z \in \mathcal{L}_2$ and $\bar{Q}^T \Gamma F_e = 0$ we conclude from (7.31) that $\bar{Q}^T \dot{\hat{\theta}} \in \mathcal{L}_1$. We now write $\bar{Q}^T \hat{\theta}(t)$ as

$$\bar{Q}^T \hat{\theta}(t) = \bar{Q}^T \hat{\theta}(0) + \int_0^t \bar{Q}^T \dot{\hat{\theta}}(\tau) d\tau. \tag{7.32}$$

The fact that $\bar{Q}^T \dot{\hat{\theta}} \in \mathcal{L}_1$ assures that $\bar{Q}^T \hat{\theta}(t) \to$ *const*. That is, we have that $[\bar{P}, \bar{Q}]^T \hat{\theta}(t) \to$ *const*. (An easy way to see this is to write $\int_0^t \bar{Q}^T \dot{\hat{\theta}}(\tau) d\tau$ as a Lebesgue integral and perform the standard decomposition $\int_{[0,t]} \bar{Q}^T \dot{\hat{\theta}} d\lambda = \int_{[0,t]} \bar{Q}^T \dot{\hat{\theta}}^+ d\lambda - \int_{[0,t]} \bar{Q}^T \dot{\hat{\theta}}^- d\lambda$; since both $\int_{[0,t]} \bar{Q}^T \dot{\hat{\theta}}^+ d\lambda$ and $\int_{[0,t]} \bar{Q}^T \dot{\hat{\theta}}^- d\lambda$ are monotonically increasing and bounded, follows that $\lim_{t \to \infty} \int_0^t \bar{Q}^T \dot{\hat{\theta}}(\tau) d\tau$..) Since $\Gamma = \Gamma^T > 0$, one easily proves that $[\bar{P}, \bar{Q}]^T$ is invertible. Thus $\hat{\theta}(t) \to$ *const*. □

An alternative proof of Theorem 7.2 for $r = 0$ provides additional insight into the mechanism of parameter convergence.

Alternative Proof of Theorem 7.2 for $r = 0$. Theorem 7.1 guarantees that $\hat{\theta}$ and z are bounded. By the Bolzano-Weierstrass theorem (see, e.g., [94]), there exists a sequence $\{t_i\}$ with $t_i \to \infty$ as $i \to \infty$, such that the sequence $\tilde{\theta}(t_i)$ has a limit $\tilde{\theta}_\infty \in \mathbb{R}^p$:

$$\tilde{\theta}(t_i) \to \tilde{\theta}_\infty \quad \text{as} \quad i \to \infty. \tag{7.33}$$

Let us introduce $\bar{\theta} = \tilde{\theta} - \tilde{\theta}_\infty = \hat{\theta}_\infty - \hat{\theta}$, and for $t \in [t_i, t_{i+1}]$ consider

$$\bar{\theta}(t) = \bar{\theta}(t_i) + \int_{t_i}^t \dot{\bar{\theta}}(\tau) d\tau. \tag{7.34}$$

In view of $|W(z(t), \tilde{\theta}(t))| \leq L|z(t)|$, we have

$$|\bar{\theta}(t)| \leq |\bar{\theta}(t_i)| + \bar{\lambda}(\Gamma) L \int_{t_i}^t |z(\tau)|^2 d\tau. \tag{7.35}$$

From (7.16) it follows that

$$\int_{t_i}^t |z(\tau)|^2 d\tau \leq -\frac{1}{c_0} \int_{t_i}^t \dot{V}_n(\tau) d\tau$$
$$= \frac{1}{c_0} (V_n(t_i) - V_n(t)), \tag{7.36}$$

and, since $\dot{V}_n \leq 0$ implies $V_n(t_{i+1}) \leq V_n(t)$, we get

$$\int_{t_i}^{t} |z(\tau)|^2 d\tau \leq \frac{1}{c_0}(V_n(t_i) - V_n(t_{i+1}))$$
$$= \frac{1}{2c_0}\left(|z(t_i)|^2 - |z(t_{i+1})|^2 + |\tilde{\theta}(t_i)|^2_{\Gamma^{-1}} - |\tilde{\theta}(t_{i+1})|^2_{\Gamma^{-1}}\right). \quad (7.37)$$

Noting that

$$|\tilde{\theta}(t_i)|^2_{\Gamma^{-1}} - |\tilde{\theta}(t_{i+1})|^2_{\Gamma^{-1}}$$
$$= \left(\tilde{\theta}(t_i) - \tilde{\theta}(t_{i+1})\right)^{\mathrm{T}} \Gamma^{-1} \left(\tilde{\theta}(t_i) + \tilde{\theta}(t_{i+1})\right)$$
$$= \left(\bar{\theta}(t_i) - \bar{\theta}(t_{i+1})\right)^{\mathrm{T}} \Gamma^{-1} \left(\bar{\theta}(t_i) + \bar{\theta}(t_{i+1}) + 2\tilde{\theta}_\infty\right)$$
$$\leq \frac{1}{\underline{\lambda}(\Gamma)} \left(|\bar{\theta}(t_i)| + |\bar{\theta}(t_{i+1})|\right) \left(|\bar{\theta}(t_i)| + |\bar{\theta}(t_{i+1})| + 2|\tilde{\theta}_\infty|\right), \quad (7.38)$$

and substituting (7.37) and (7.38) into (7.35), we get

$$|\bar{\theta}(t)| \leq |\bar{\theta}(t_i)| + \frac{\bar{\lambda}(\Gamma) L}{2c_0}\Big[|z(t_i)|^2$$
$$+ \frac{1}{\underline{\lambda}(\Gamma)}\left(|\bar{\theta}(t_i)| + |\bar{\theta}(t_{i+1})|\right)\left(|\bar{\theta}(t_i)| + |\bar{\theta}(t_{i+1})| + 2|\tilde{\theta}_\infty|\right)\Big], \quad (7.39)$$

where $\underline{\lambda}(\Gamma)$ and $\bar{\lambda}(\Gamma)$ respectively denote the smallest and the largest eigenvalue of Γ. Now we let $i \to \infty$, which means that $t \in [t_i, t_{i+1}]$ also tends to ∞. Since Theorem 7.1 established that $z(t) \to 0$ as $t \to \infty$, and because of (7.33), the right-hand side of (7.39) tends to zero, which means that $\bar{\theta}(t) \to 0$, i.e., $\hat{\theta}(t) \to \hat{\theta}_\infty$. □

7.3 Decomposition of the Parameter Vector

We first prove the following lemmas.

Lemma 7.3 *Let $F_e^{\mathrm{T}}\tilde{\theta} = 0$. Then $z = 0$ if and only if $x = x_e$.*

Proof. We start by noting that $z_1 = 0$ iff $x_1 = x_1^e$. Assume that $z_k = 0$ iff $x_k = x_k^e$, $k = 1, \cdots, i-1$. Recalling that

$$\alpha_k = -z_{k-1} - c_k z_k - \varphi_k(\bar{x}_k)^{\mathrm{T}}\hat{\theta} + \sum_{j=1}^{k-1} \frac{\partial \alpha_{k-1}}{\partial x_j}\left(x_{j+1} + \varphi_j(\bar{x}_j)^{\mathrm{T}}\hat{\theta}\right)$$
$$+ \frac{\partial \alpha_{k-1}}{\partial \hat{\theta}}\Gamma \tau_k - \sum_{j=2}^{k-1} \sigma_{j,k} z_k, \quad (7.40)$$

7.3 DECOMPOSITION OF THE PARAMETER VECTOR

we have

$$\begin{aligned}
z_i &= x_i - \alpha_{i-1} \\
&= x_i - x_i^e - \varphi_{i-1}(\bar{x}_{i-1}^e)^T\theta - \alpha_{i-1} \\
&= x_i - x_i^e - \underbrace{\varphi_{i-1}(\bar{x}_{i-1}^e)^T\tilde{\theta}}_{0} + \underbrace{\left(\varphi_{i-1}(\bar{x}_{i-1}) - \varphi_{i-1}(\bar{x}_{i-1}^e)\right)^T\hat{\theta}}_{0} \\
&\quad - \sum_{k=1}^{i-2} \frac{\partial \alpha_{i-2}}{\partial x_k}\underbrace{(\varphi_k(\bar{x}_k) - \varphi_k(\bar{x}_k^e))^T\hat{\theta}}_{0} - \sum_{k=1}^{i-2} \frac{\partial \alpha_{i-2}}{\partial x_k}\underbrace{(x_{k+1} - x_{k+1}^e)}_{0} \\
&\quad + \underbrace{z_{i-2}}_{0} + c_{i-1}\underbrace{z_{i-1}}_{0} - \frac{\partial \alpha_{i-2}}{\partial \hat{\theta}}\Gamma \underbrace{\tau_{i-1}}_{0} + \sum_{k=2}^{i-2} \sigma_{k,i-1}\underbrace{z_{i-1}}_{0} \\
&= x_i - x_i^e.
\end{aligned} \tag{7.41}$$

By induction, we conclude that $z = 0$ iff $x = x^e$. □

Since each solution of the adaptive system converges to an equilibrium point on M, it is first of interest to determine which of the equilibria on M are stable, and which are unstable. Let us for further notational convenience rewrite the system (7.10) and (7.11) as

$$\dot{z} = A_z(z,\tilde{\vartheta})z + \mathcal{W}(z,\tilde{\vartheta})^T\tilde{\vartheta} \tag{7.42}$$

$$\dot{\tilde{\vartheta}} = -\Gamma \mathcal{W}(z,\tilde{\vartheta})z \tag{7.43}$$

and denote $\Gamma = \gamma\Gamma_0$, where $\gamma = \lambda_{max}(\Gamma)$. Then we have the following lemma.

Lemma 7.4 *The system (7.42) and (7.43) is transformed into*

$$\dot{z} = A_z(z,\tilde{\theta}_1,\tilde{\theta}_2)z + W_1(z,\tilde{\theta}_1,\tilde{\theta}_2)^T\tilde{\theta}_1 + W_2(z,\tilde{\theta}_1,\tilde{\theta}_2)^T\tilde{\theta}_2 \tag{7.44}$$

$$\dot{\tilde{\theta}}_1 = -\gamma W_1(z,\tilde{\theta}_1,\tilde{\theta}_2)z \tag{7.45}$$

$$\dot{\tilde{\theta}}_2 = -\gamma W_2(z,\tilde{\theta}_1,\tilde{\theta}_2)z \tag{7.46}$$

using the transformation

$$\begin{aligned}
\tilde{\theta} &= \begin{bmatrix} \tilde{\theta}_1 \\ \tilde{\theta}_2 \end{bmatrix} = T\tilde{\vartheta} \\
T &= \begin{bmatrix} (P^TP)^{-1/2}P^T \\ (Q^TQ)^{-1/2}Q^T \end{bmatrix} \Gamma_0^{-1/2} \\
T^{-1} &= \Gamma_0^{1/2} \begin{bmatrix} P(P^TP)^{-1/2} & Q(Q^TQ)^{-1/2} \end{bmatrix}
\end{aligned} \tag{7.47}$$

$P =$ *matrix of basis vectors of* $\text{Range}\{\Gamma_0^{1/2}F_e\}$

$Q =$ *matrix of basis vectors of* $\text{Null}\{(\Gamma_0^{1/2}F_e)^T\}$.

Furthermore, $\text{rank}\{W_1(0,0,\tilde{\theta}_2)\} = r$ *and* $W_2(0,0,\tilde{\theta}_2) = 0$ *for all* $\tilde{\theta}_2 \in \mathbb{R}^{p-r}$.

This decomposition clearly separates the part of the parameter error vector which is guaranteed to converge to a zero vector from the part which converges to a possibly nonzero constant vector. To see this, recall that $F_e^T \tilde{\vartheta}(t) \to 0$ which implies that $F_e^T \Gamma_0^{1/2} P(P^T P)^{-1/2} \tilde{\theta}_1(t) \to 0$. Since $\text{rank}\{F_e^T \Gamma_0^{1/2} P(P^T P)^{-1/2}\} = \deg\{\tilde{\theta}_1\}$, then $\tilde{\theta}_1(t) \to 0$.

Proof of Lemma 7.4. Applying the transformation (7.47), the system (7.42), (7.43) becomes

$$\begin{aligned}
\dot{z} &= A_z(z, \tilde{\vartheta})z + W(z, \tilde{\vartheta})^T \tilde{\vartheta} \\
&= A_z(z, T^{-1}\tilde{\theta})z + NF^T T^{-1} T \tilde{\vartheta} \\
&= A_z(z, \tilde{\theta}_1, \tilde{\theta}_2)z + NF^T \Gamma_0^{1/2} \begin{bmatrix} P(P^T P)^{-1/2} & Q(Q^T Q)^{-1/2} \end{bmatrix} \tilde{\theta} \\
&= A_z(z, \tilde{\theta}_1, \tilde{\theta}_2)z + W_1(z, \tilde{\theta}_1, \tilde{\theta}_2)^T \tilde{\theta}_1 + W_2(z, \tilde{\theta}_1, \tilde{\theta}_2)^T \tilde{\theta}_2 \quad (7.48)
\end{aligned}$$

$$\begin{aligned}
\dot{\tilde{\theta}} &= \begin{bmatrix} \dot{\tilde{\theta}}_1 \\ \dot{\tilde{\theta}}_2 \end{bmatrix} = -\gamma \begin{bmatrix} (P^T P)^{-1/2} P^T \\ (Q^T Q)^{-1/2} Q^T \end{bmatrix} \Gamma_0^{1/2} F N^T z \\
&= \begin{bmatrix} -\gamma W_1(z, \tilde{\theta}_1, \tilde{\theta}_2) z \\ -\gamma W_2(z, \tilde{\theta}_1, \tilde{\theta}_2) z \end{bmatrix}, \quad (7.49)
\end{aligned}$$

where

$$W_1(z, \tilde{\theta}_1, \tilde{\theta}_2) = (P^T P)^{-1/2} P^T \Gamma_0^{1/2} F N^T \quad (7.50)$$
$$W_2(z, \tilde{\theta}_1, \tilde{\theta}_2) = (Q^T Q)^{-1/2} Q^T \Gamma_0^{1/2} F N^T. \quad (7.51)$$

Using Lemma 7.3, we can see that $\tilde{\theta}_1 = 0$ and $z = 0$ imply $x = x_e$. Thus (7.50) and (7.51) yield

$$W_1(0, 0, \tilde{\theta}_2) = (P^T P)^{-1/2} P^T \Gamma_0^{1/2} F_e N^T \big|_{z, \tilde{\theta}_1 = 0} \quad (7.52)$$
$$W_2(0, 0, \tilde{\theta}_2) = (Q^T Q)^{-1/2} Q^T \Gamma_0^{1/2} F_e N^T \big|_{z, \tilde{\theta}_1 = 0}. \quad (7.53)$$

Since $\text{rank}\{(P^T P)^{-1/2} P^T \Gamma_0^{1/2} F_e\} = r \leq n$ and $\text{rank}\{N\} = n$ for all $\tilde{\theta}_2$, then $\text{rank}\{W_1(0, 0, \tilde{\theta}_2)\} \equiv r$. On the other hand, since $Q^T \Gamma_0^{1/2} F_e = 0$, then $W_2(0, 0, \tilde{\theta}_2) \equiv 0$. □

7.4 Center Manifold Analysis

Since all the solutions converge to the $\tilde{\theta}_2$-subspace, that is, the manifold

$$M = \{(z, \tilde{\theta}_1, \tilde{\theta}_2) \in \mathbb{R}^{n+p} | z = 0, \tilde{\theta}_1 = 0\}, \quad (7.54)$$

7.4 CENTER MANIFOLD ANALYSIS

we study stability of equilibria on M as a function of $\tilde{\theta}$. Let us denote

$$A_e(\tilde{\theta}_2) = \begin{bmatrix} A_0 + \gamma\Sigma(\tilde{\theta}_2) + \dfrac{\partial W_2(0,0,\tilde{\theta}_2)^T \tilde{\theta}_2}{\partial z} & W_1^T(0,0,\tilde{\theta}_2) + \dfrac{\partial W_2(0,0,\tilde{\theta}_2)^T \tilde{\theta}_2}{\partial \tilde{\theta}_1} \\ -\gamma W_1(0,0,\tilde{\theta}_2) & 0 \end{bmatrix},$$
(7.55)

where

$$A_0 = \begin{bmatrix} -c_1 & 1 & & & \\ -1 & \ddots & \ddots & & \\ & \ddots & \ddots & 1 & \\ & & -1 & -c_n \end{bmatrix}$$
(7.56)

$$\Sigma(\tilde{\theta}_2) = \begin{bmatrix} 0 & 0 & 0 & \cdots & 0 \\ 0 & 0 & \bar{\sigma}_{23} & \cdots & \bar{\sigma}_{2n} \\ 0 & -\bar{\sigma}_{23} & \ddots & \ddots & \vdots \\ \vdots & \vdots & \ddots & \ddots & \bar{\sigma}_{n-1,n} \\ 0 & -\bar{\sigma}_{2n} & \cdots & -\bar{\sigma}_{n-1,n} & 0 \end{bmatrix}$$
(7.57)

$$\bar{\sigma}_{jk}(\tilde{\theta}_2) = -\dfrac{\partial \alpha_{j-1}}{\partial \hat{\vartheta}} \Gamma_0 w_k \bigg|_{\substack{z=0 \\ \tilde{\theta}_1 = 0}} \qquad k = 1,\cdots,n.$$
(7.58)

We first prove a technical lemma which prepares the system for center manifold analysis.

Lemma 7.5 *The system (7.44)-(7.46) can be rewritten as*

$$\begin{bmatrix} \dot{z} \\ \dot{\tilde{\theta}}_1 \end{bmatrix} = A_e(\tilde{\theta}_2^e) \begin{bmatrix} z \\ \tilde{\theta}_1 \end{bmatrix} + \begin{bmatrix} G(z,\tilde{\theta}_1,\bar{\theta}_2) \\ G_5(z,\tilde{\theta}_1,\bar{\theta}_2) \end{bmatrix}$$
$$\dot{\bar{\theta}}_2 = H(z,\tilde{\theta}_1,\bar{\theta}_2),$$
(7.59)

where $\bar{\theta}_2 = \tilde{\theta}_2 - \tilde{\theta}_2^e$, and $G(z,\tilde{\theta}_1,\bar{\theta}_2)$ and $H(z,\tilde{\theta}_1,\bar{\theta}_2)$ satisfy

$$G(0,0,\bar{\theta}_2) = 0, \quad \dfrac{\partial G(0,0,0)}{\partial z} = 0, \quad \dfrac{\partial G(0,0,0)}{\partial \tilde{\theta}_1} = 0, \quad \dfrac{\partial G(0,0,\bar{\theta}_2)}{\partial \bar{\theta}_2} = 0$$

$$G_5(0,0,\bar{\theta}_2) = 0, \quad \dfrac{\partial G_5(0,0,0)}{\partial z} = 0, \quad \dfrac{\partial G_5(0,0,0)}{\partial \tilde{\theta}_1} = 0, \quad \dfrac{\partial G_5(0,0,\bar{\theta}_2)}{\partial \bar{\theta}_2} = 0$$

$$H(0,0,\bar{\theta}_2) = 0, \quad \dfrac{\partial H(0,0,\bar{\theta}_2)}{\partial z} = 0, \quad \dfrac{\partial H(0,0,\bar{\theta}_2)}{\partial \tilde{\theta}_1} = 0, \quad \dfrac{\partial H(0,0,\bar{\theta}_2)}{\partial \bar{\theta}_2} = 0 \;.$$
(7.60)

Proof. Noting that $A_z(z,\tilde{\theta}_1,\tilde{\theta}_2)z = A_z(0,0,\tilde{\theta}_2^e)z + \left(A_z(z,\tilde{\theta}_1,\tilde{\theta}_2) - A_z(0,0,\tilde{\theta}_2^e)\right)z$, there exists a smooth vector valued function $G_1(z,\tilde{\theta}_1,\tilde{\theta}_2)$, with

$G_1(0, \tilde{\theta}_1, \tilde{\theta}_2) \equiv 0$, $\dfrac{\partial G_1(0,0,\tilde{\theta}_2)}{\partial z} \equiv 0$, $\dfrac{\partial G_1(0,\tilde{\theta}_1,\tilde{\theta}_2)}{\partial \tilde{\theta}_1} \equiv 0$ and $\dfrac{\partial G_1(0,\tilde{\theta}_1,\tilde{\theta}_2)}{\partial \tilde{\theta}_2} \equiv 0$, such that

$$A_z(z, \tilde{\theta}_1, \tilde{\theta}_2) z = A_z(0, 0, \tilde{\theta}_2^e) z + G_1(z, \tilde{\theta}_1, \tilde{\theta}_2). \tag{7.61}$$

From (7.12) and (7.56)–(7.58) it follows that $A_z(0,0,\tilde{\theta}_2^e) = A_0 + \gamma \Sigma(\tilde{\theta}_2^e)$, so we have

$$A_z(z, \tilde{\theta}_1, \tilde{\theta}_2) z = A_0 + \gamma \Sigma(\tilde{\theta}_2^e) + G_1(z, \tilde{\theta}_1, \tilde{\theta}_2). \tag{7.62}$$

Since $W_1(z, \tilde{\theta}_1, \tilde{\theta}_2)^T \tilde{\theta}_1 = W_1(0,0,\tilde{\theta}_2^e)^T \tilde{\theta}_1 + \left(W_1(z,\tilde{\theta}_1,\tilde{\theta}_2) - W_1(0,0,\tilde{\theta}_2^e)\right)^T \tilde{\theta}_1$, there exists a smooth vector valued function $G_2(z, \tilde{\theta}_1, \tilde{\theta}_2)$, with $G_2(z, 0, \tilde{\theta}_2) \equiv 0$, $\dfrac{\partial G_2(z,0,\tilde{\theta}_2)}{\partial z} \equiv 0$, $\dfrac{\partial G_2(0,0,\tilde{\theta}_2)}{\partial \tilde{\theta}_1} \equiv 0$, and $\dfrac{\partial G_2(z,0,\tilde{\theta}_2)}{\partial \tilde{\theta}_2} \equiv 0$, such that

$$W_1(z, \tilde{\theta}_1, \tilde{\theta}_2)^T \tilde{\theta}_1 = W_1(0,0,\tilde{\theta}_2^e)^T \tilde{\theta}_1 + G_2(z, \tilde{\theta}_1, \tilde{\theta}_2). \tag{7.63}$$

Since $W_2(0, 0, \tilde{\theta}_2) \equiv 0$, there exists a smooth vector valued function $G_3(z, \tilde{\theta}_1)$, with $G_3(0,0) \equiv 0$, $\dfrac{\partial G_3(0,0)}{\partial z} \equiv 0$ and $\dfrac{\partial G_3(0,0)}{\partial \tilde{\theta}_1} \equiv 0$, such that

$$\begin{aligned}
W_2(z, \tilde{\theta}_1, \tilde{\theta}_2)^T \tilde{\theta}_2 &= W_2(z, \tilde{\theta}_1, \tilde{\theta}_2^e)^T \tilde{\theta}_2^e + W_2(z, \tilde{\theta}_1, \tilde{\theta}_2)^T \tilde{\theta}_2 - W_2(z, \tilde{\theta}_1, \tilde{\theta}_2^e)^T \tilde{\theta}_2^e \\
&= \dfrac{\partial W_2(0,0,\tilde{\theta}_2^e)^T \tilde{\theta}_2^e}{\partial z} z + \dfrac{\partial W_2(0,0,\tilde{\theta}_2^e)^T \tilde{\theta}_2^e}{\partial \tilde{\theta}_1} \tilde{\theta}_1 + G_3(z, \tilde{\theta}_1) \\
&\quad + W_2(z, \tilde{\theta}_1, \tilde{\theta}_2)^T \tilde{\theta}_2 - W_2(z, \tilde{\theta}_1, \tilde{\theta}_2^e)^T \tilde{\theta}_2^e.
\end{aligned} \tag{7.64}$$

Introducing a new variable $\bar{\theta}_2 = \tilde{\theta}_2 - \tilde{\theta}_2^e$, we conclude that there exists a smooth vector valued function $G_4(z, \tilde{\theta}_1, \bar{\theta}_2)$, with $G_4(0, 0, \bar{\theta}_2) = G_4(z, \tilde{\theta}_1, 0) \equiv 0$, $\dfrac{\partial G_4(z,\tilde{\theta}_1,0)}{\partial z} \equiv 0$, $\dfrac{\partial G_4(z,\tilde{\theta}_1,0)}{\partial \tilde{\theta}_1} \equiv 0$ and $\dfrac{\partial G_4(0,0,\bar{\theta}_2)}{\partial \bar{\theta}_2} \equiv 0$, such that

$$\begin{aligned}
W_2(z, \tilde{\theta}_1, \tilde{\theta}_2)^T \tilde{\theta}_2 &= \dfrac{\partial W_2(0,0,\tilde{\theta}_2^e)^T \tilde{\theta}_2^e}{\partial z} z + \dfrac{\partial W_2(0,0,\tilde{\theta}_2^e)^T \tilde{\theta}_2^e}{\partial \tilde{\theta}_1} \tilde{\theta}_1 + G_3(z, \tilde{\theta}_1) \\
&\quad + G_4(z, \tilde{\theta}_1, \bar{\theta}_2).
\end{aligned} \tag{7.65}$$

Now since $W_1(z,\tilde{\theta}_1,\tilde{\theta}_2)z = W_1(z,\tilde{\theta}_1,\tilde{\theta}_2^e)z + W_1(z,\tilde{\theta}_1,\tilde{\theta}_2)z - W_1(z,\tilde{\theta}_1,\tilde{\theta}_2^e)z$, there exists $G_5(z, \tilde{\theta}_1, \bar{\theta}_2)$ with $G_5(0, \tilde{\theta}_1, \bar{\theta}_2) \equiv 0$, $\dfrac{\partial G_5(0,0,\bar{\theta}_2)}{\partial z} \equiv 0$, $\dfrac{\partial G_5(0,\tilde{\theta}_1,\bar{\theta}_2)}{\partial \tilde{\theta}_1} \equiv 0$ and $\dfrac{\partial G_5(0,\tilde{\theta}_1,\bar{\theta}_2)}{\partial \bar{\theta}_2} \equiv 0$, such that

$$-\gamma W_1(z, \tilde{\theta}_1, \tilde{\theta}_2) z = -\gamma W_1(z, \tilde{\theta}_1, \tilde{\theta}_2^e) z + G_5(z, \tilde{\theta}_1, \bar{\theta}_2). \tag{7.66}$$

7.4 CENTER MANIFOLD ANALYSIS

Since $W_2(0,0,\bar{\theta}_2) \equiv 0$, there exists $H(z,\tilde{\theta}_1,\bar{\theta}_2)$ with $H(0,\tilde{\theta}_1,\bar{\theta}_2) \equiv 0$, $\dfrac{\partial H(0,0,\bar{\theta}_2)}{\partial z} \equiv 0$, $\dfrac{\partial H(0,\tilde{\theta}_1,\bar{\theta}_2)}{\partial \tilde{\theta}_1} \equiv 0$ and $\dfrac{\partial H(0,0,\bar{\theta}_2)}{\partial \bar{\theta}_2} \equiv 0$, such that

$$-\gamma W_2(z,\tilde{\theta}_1,\bar{\theta}_2)z = H(z,\tilde{\theta}_1,\bar{\theta}_2). \tag{7.67}$$

Denoting

$$G(z,\tilde{\theta}_1,\bar{\theta}_2) = G_1(z,\tilde{\theta}_1,\tilde{\theta}_2^e + \bar{\theta}_2) + G_2(z,\tilde{\theta}_1,\tilde{\theta}_2^e + \bar{\theta}_2) + G_3(z,\tilde{\theta}_1) + G_4(z,\tilde{\theta}_1,\bar{\theta}_2), \tag{7.68}$$

in view of (7.62), (7.63), (7.65), (7.66), and (7.67), we obtain (7.59). The properties of G_1, G_2, G_3, G_4, G_5, and H imply (7.60). □

We are now ready to start the categorization of the equilibria into the stable and the unstable ones.

Theorem 7.6 *Consider the closed-loop adaptive system (7.44), (7.45), (7.46). The equilibrium $(z,\tilde{\theta}_1,\tilde{\theta}_2) = (0,0,\tilde{\theta}_2^e)$ is*

1. *globally stable if all the eigenvalues of $A_e(\tilde{\theta}_2^e)$ have negative real parts;*

2. *unstable if at least one eigenvalue of $A_e(\tilde{\theta}_2^e)$ has positive real part.*

Proof. Using Lemma 7.5, we transform system (7.44)–(7.46) into the form

$$\begin{bmatrix} \dot{z} \\ \dot{\tilde{\theta}}_1 \end{bmatrix} = A_e(\tilde{\theta}_2^e)\begin{bmatrix} z \\ \tilde{\theta}_1 \end{bmatrix} + \begin{bmatrix} G(z,\tilde{\theta}_1,\bar{\theta}_2) \\ G_5(z,\tilde{\theta}_1,\bar{\theta}_2) \end{bmatrix} \tag{7.69}$$

$$\dot{\bar{\theta}}_2 = H(z,\tilde{\theta}_1,\bar{\theta}_2),$$

where $\bar{\theta}_2 = \tilde{\theta}_2 - \tilde{\theta}_2^e$, and $G(z,\tilde{\theta}_1,\bar{\theta}_2)$ and $H(z,\tilde{\theta}_1,\bar{\theta}_2)$ satisfy

$$G(0,\bar{\theta}_2) = 0, \quad \frac{\partial G(0)}{\partial z} = 0, \quad \frac{\partial G(0)}{\partial \tilde{\theta}_1} = 0, \quad \frac{\partial G(0,\bar{\theta}_2)}{\partial \bar{\theta}_2} = 0$$

$$G_5(0,\bar{\theta}_2) = 0, \quad \frac{\partial G_5(0)}{\partial z} = 0, \quad \frac{\partial G_5(0)}{\partial \tilde{\theta}_1} = 0, \quad \frac{\partial G_5(0,\bar{\theta}_2)}{\partial \bar{\theta}_2} = 0 \tag{7.70}$$

$$H(0,\bar{\theta}_2) = 0, \quad \frac{\partial H(0,\bar{\theta}_2)}{\partial z} = 0, \quad \frac{\partial H(0,\bar{\theta}_2)}{\partial \tilde{\theta}_1} = 0, \quad \frac{\partial H(0,\bar{\theta}_2)}{\partial \bar{\theta}_2} = 0 \ .$$

We first prove the stability part, assuming that all the eigenvalues of $A_e(\tilde{\theta}_2^e)$ have negative real parts. Since the equilibrium manifold $[z^T,\tilde{\theta}_1^T]^T = h(\bar{\theta}_2) = 0$ is invariant and $\dfrac{\partial h(0)}{\partial \bar{\theta}_2} = 0$, then $[z^T,\tilde{\theta}_1^T]^T = 0$ is a center manifold. The reduced system of system (7.69),

$$\dot{\bar{\theta}}_2 = H(0,0,\bar{\theta}_2) = 0, \tag{7.71}$$

is stable. By the center manifold theorem (reduction principle) [12, Theorem 2 on p. 21], the equilibrium $(z, \tilde{\theta}_1, \bar{\theta}_2) = (0,0,0)$ of the full system (7.69) is stable. The stability property is global because Theorem 7.1 guarantees global boundedness.

The instability part, when at least one of the eigenvalues of $A_e(\tilde{\theta}_2^e)$ has positive real part, is immediate from the linearization theorem by noting that the linearization of (7.69) around the equilibrium $(z, \tilde{\theta}_1, \bar{\theta}_2) = (0,0,0)$ is

$$\begin{bmatrix} \delta \dot{z} \\ \delta \dot{\tilde{\theta}}_1 \\ \delta \dot{\bar{\theta}}_2 \end{bmatrix} = \begin{bmatrix} A_e(\tilde{\theta}_2^e) & 0 & 0 \\ 0 & 0 & 0 \\ 0 & 0 & 0 \end{bmatrix} \begin{bmatrix} \delta z \\ \delta \tilde{\theta}_1 \\ \delta \bar{\theta}_2 \end{bmatrix}. \tag{7.72}$$

□

Remark 7.7 (critical case) Theorem 7.6 does not cover the case where, in addition to eigenvalues with nonnegative real parts, there are also eigenvalues with zero real parts. For example, consider the case with $r = 0$. In this case, the z-system in (7.69) has to be decomposed as

$$\begin{aligned} \dot{z}^s &= A_s(\tilde{\theta}^e)z^s + G_s(z^s, z^c, \bar{\theta}) \\ \dot{z}^c &= A_c(\tilde{\theta}^e)z^c + G_c(z^s, z^c, \bar{\theta}) \\ \dot{\bar{\theta}} &= H'(z^s, z^c, \bar{\theta}), \end{aligned} \tag{7.73}$$

where $A_s(\tilde{\theta}^e)$ has all eigenvalues with negative real parts, and $A_c(\tilde{\theta}^e)$ has all eigenvalues with zero real parts. Let us denote by E^c the (generalized) eigenspace corresponding to *all* the eigenvalues with zero real parts of the matrix diag$\{A_s, A_c, 0\}$ (including p eigenvalues equal to zero). Then there exists locally an invariant center manifold $W_{\text{loc}}^c(0)$ tangent to E^c at $(z^s, z^c, \bar{\theta}) = 0$, and given by $z^s = h(z^c, \bar{\theta})$, with $h(0,0) = 0$, $\frac{\partial h(0,0)}{\partial z^c} = 0$, $\frac{\partial h(0,0)}{\partial \theta} = 0$. To determine whether the equilibrium $(z^s, z^c, \bar{\theta}) = 0$ is stable, we need to see if the equilibrium $(z^c, \bar{\theta}) = 0$ of the reduced system

$$\begin{aligned} \dot{z}^c &= A_c(\tilde{\theta}^e)z^c + G_c(h(z^c, \bar{\theta}), z^c, \bar{\theta}) \\ \dot{\bar{\theta}} &= H'(h(z^c, \bar{\theta}), z^c, \bar{\theta}) \end{aligned} \tag{7.74}$$

is stable. It is important to realize two possible difficulties. While, in the case where all the eigenvalues of $A_e(\tilde{\theta}^e)$ had nonzero real parts, the center manifold was simply the equilibrium manifold $z = 0$, in the case where $A_e(\tilde{\theta}^e)$ has eigenvalues with zero real parts, the manifold $z^s = h(z^c, \bar{\theta})$ has to be separately determined. To do this, one, in general, has to use the approximation technique of [12, Theorem 3 on p. 25]. After finding the center manifold, a much more difficult task follows — checking stability of the reduced system (7.74). □

The following simple example illustrates the stability analysis at a critical point where Theorem 7.6 fails to provide an answer. The example is contrived so that approximate computation of the center manifold is avoided.

7.5 CATEGORIZATION OF EQUILIBRIA

Example 7.8 Consider the scalar system $\dot{x} = \varphi(x)\theta + u$, for which we design the adaptive controller $u = -x - \varphi(x)\hat{\theta}$ and the update law $\dot{\hat{\theta}} = \frac{1}{8}x\varphi(x)$. The closed-loop system is

$$\begin{aligned}\dot{x} &= -x + \varphi(x)\tilde{\theta} \\ \dot{\tilde{\theta}} &= -\tfrac{1}{8}x\varphi(x).\end{aligned} \qquad (7.75)$$

We consider two cases: $\varphi(x) = x + |x|x$ and $\varphi(x) = x - |x|x$. In both cases, Theorem 7.6 establishes that the equilibria on the interval $(-1+\theta, +\infty)$ of the $\hat{\theta}$-axis are stable, while those on the interval $(-\infty, -1+\theta)$ are unstable. It does not say anything, however, about the stability of the equilibrium $(x, \hat{\theta}) = (0, -1 + \theta)$. Introducing the new variable $\bar{\theta} = \tilde{\theta} - 1 = -1 + \theta - \hat{\theta}$, we rewrite (7.75) as

$$\begin{aligned}\dot{x} &= \varphi(x)\bar{\theta} + (\varphi(x) - x) \\ \dot{\bar{\theta}} &= -\tfrac{1}{8}x\varphi(x).\end{aligned} \qquad (7.76)$$

With the Lyapunov function $V = x^2 + 8\bar{\theta}^2$, along the solutions of (7.76), we have $\dot{V} = 2x(\varphi(x) - x)$. In the case $\varphi(x) = x - |x|x$, the Lyapunov derivative becomes

$$\dot{V} = -2|x|^3 \leq 0, \qquad (7.77)$$

and it follows that the equilibrium $(x, \hat{\theta}) = (0, -1 + \theta)$ is globally stable. However, in the case $\varphi(x) = x + |x|x$, the Lyapunov derivative is

$$\dot{V} = 2|x|^3 \geq 0. \qquad (7.78)$$

While one might be tempted to conclude that the equilibrium $(x, \hat{\theta}) = (0, -1+\theta)$ is unstable, (7.78) is not enough as it does not satisfy all the conditions of Chetaev's theorem. Consider instead the function

$$U = x^2 - 4\bar{\theta}^2. \qquad (7.79)$$

Lengthy calculation shows that

$$\dot{U} \geq \frac{1}{4}|x|^3 + \frac{3}{2}x^2(|x| - 2|\bar{\theta}|) + \frac{1}{4}|x|^3 \left(1 - 12\sqrt{x^2 + \bar{\theta}^2}\right), \qquad (7.80)$$

which proves that $\dot{U} > 0$ on the set

$$\Omega = \left\{12\sqrt{x^2 + \bar{\theta}^2} < 1 \mid U(x, \bar{\theta}) > 0\right\}. \qquad (7.81)$$

Conditions of Chetaev's theorem (see, e.g., [58, Theorem 3.3]) are satisfied, which proves that the equilibrium $(x, \hat{\theta}) = (0, -1 + \theta)$ is unstable. □

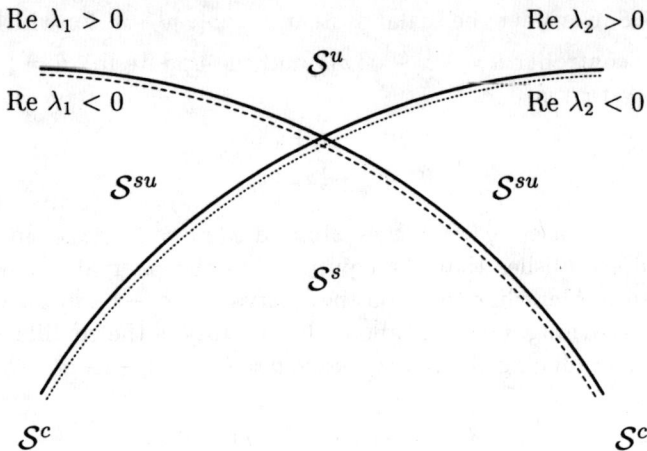

Figure 7.1: Example of sets $\mathcal{S}^s, \mathcal{S}^u, \mathcal{S}^{su}, \mathcal{S}^c$, in the $\hat{\theta}$-plane for $n = p = 2$.

7.5 Categorization of Equilibria

Let us now categorize the equilibria on M into the following four sets:

$$\mathcal{S}^s = \bigcap_{i=1}^{n} \left\{ (z, \tilde{\theta}_1, \tilde{\theta}_2) \in M \mid \text{Re } \lambda_i(A_e(\tilde{\theta}_2)) < 0 \right\} \quad (7.82)$$

$$\mathcal{S}^u = \bigcap_{i=1}^{n} \left\{ (z, \tilde{\theta}_1, \tilde{\theta}_2) \in M \mid \text{Re } \lambda_i(A_e(\tilde{\theta}_2)) > 0 \right\} \quad (7.83)$$

$$\mathcal{S}^c = \bigcup_{i=1}^{n} \left\{ (z, \tilde{\theta}_1, \tilde{\theta}_2) \in M \mid \text{Re } \lambda_i(A_e(\tilde{\theta}_2)) = 0 \right\} \quad (7.84)$$

$$\mathcal{S}^{su} = M \setminus (\mathcal{S}^s \cup \mathcal{S}^u \cup \mathcal{S}^c). \quad (7.85)$$

An example of these sets for $n = p = 2$ is given in Figure 7.1. The set $\mathcal{S}^s$ is, by Theorem 7.6, a set of stable equilibria. The set $\mathcal{S}^u \cup \mathcal{S}^{su}$ is the set of unstable equilibria from the second part of Theorem 7.6. The set $\mathcal{S}^c$ is a set of equilibria at which at least one of the eigenvalues of $A_e(\tilde{\theta}_2)$ has zero real part. Each of the sets $\mathcal{S}^s, \mathcal{S}^u, \mathcal{S}^{su}$ is open in M, and its boundary is smooth. The set $\mathcal{S}^c$ is the boundary of $\mathcal{S}^s \cup \mathcal{S}^u \cup \mathcal{S}^{su}$. Since $A_e(\tilde{\theta}_2)$ depends on θ, the designer cannot a priori determine $\mathcal{S}^s, \mathcal{S}^u, \mathcal{S}^{su}, \mathcal{S}^c$.

We now illustrate Theorem 7.6 and the categorization (7.82)–(7.85) with the following examples.

Example 7.9 Consider the second order system with an unknown scalar parameter:

$$\begin{aligned} \dot{x}_1 &= x_2 + x_1 \theta \\ \dot{x}_2 &= u. \end{aligned} \quad (7.86)$$

7.5 CATEGORIZATION OF EQUILIBRIA

In this system, $r = 0$. After the tuning function design from Section 7.1 is applied, the resulting error system is:

$$\dot{z} = \begin{bmatrix} -c_1 & 1 \\ -1 & -c_2 \end{bmatrix} z + \begin{bmatrix} 1 \\ c_1 + \hat{\theta} \end{bmatrix} z_1 \tilde{\theta}$$
$$\dot{\hat{\theta}} = -\gamma z_1 \begin{bmatrix} 1, & c_1 + \hat{\theta} \end{bmatrix} z. \tag{7.87}$$

where $z_1 = x_1$, $z_2 = x_2 + (c_1 + \hat{\theta})z_1$. For this system we have

$$A_e(\tilde{\theta}) = \begin{bmatrix} -c_1 + \tilde{\theta} & 1 \\ -1 + (c_1 + \hat{\theta})\tilde{\theta} & -c_2 \end{bmatrix}. \tag{7.88}$$

Let $c_1 = c_2 = 1$ and $\theta = 2$. The characteristic polynomial of $A_e(\hat{\theta})$ is

$$p(s) = s^2 + \hat{\theta}s + \hat{\theta}^2 - 2. \tag{7.89}$$

which gives

$$\begin{aligned} \mathcal{S}^s &= \left\{ (z, \tilde{\theta}) \in M \mid \hat{\theta} \in \left(-\infty, 2 - \sqrt{2}\right) \right\} \\ \mathcal{S}^{su} &= \left\{ (z, \tilde{\theta}) \in M \mid \hat{\theta} \in \left(2 - \sqrt{2}, 2 + \sqrt{2}\right) \right\} \\ \mathcal{S}^u &= \left\{ (z, \tilde{\theta}) \in M \mid \hat{\theta} \in \left(2 + \sqrt{2}, +\infty\right) \right\} \\ \mathcal{S}^c &= \left\{ (z, \tilde{\theta}) \in M \mid \hat{\theta} \in \left\{-\sqrt{2}, \sqrt{2}\right\} \right\}, \end{aligned} \tag{7.90}$$

where M is the $\tilde{\theta}$-axis. These sets are shown in Figure 7.2. □

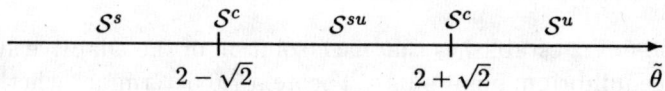

Figure 7.2: Categorization of equilibria on M into sets $\mathcal{S}^s, \mathcal{S}^u, \mathcal{S}^{su}, \mathcal{S}^c$.

Example 7.10 Let us now consider the system

$$\begin{aligned} \dot{x}_1 &= x_2 + x_1 \theta_2 \\ \dot{x}_2 &= u + \theta_1. \end{aligned} \tag{7.91}$$

In this system, $0 < r = 1 < p = 2$. After the tuning functions design from Section 7.1 is applied, the resulting error system is

$$\begin{aligned} \dot{z} &= \begin{bmatrix} -c_1 & 1 \\ -1 & -c_2 \end{bmatrix} z + \begin{bmatrix} 0 \\ 1 \end{bmatrix} \tilde{\theta}_1 + \begin{bmatrix} 1 \\ c_1 + \hat{\theta}_2 \end{bmatrix} z_1 \tilde{\theta}_2 \\ \dot{\hat{\theta}}_1 &= -\gamma \begin{bmatrix} 0 & 1 \end{bmatrix} z \\ \dot{\hat{\theta}}_2 &= -\gamma z_1 \begin{bmatrix} 1 & c_1 + \hat{\theta}_2 \end{bmatrix} z, \end{aligned} \tag{7.92}$$

where $z_1 = x_1$, $z_2 = x_2 + (c_1 + \hat{\theta}_2)z_1$. Therefore

$$A_e(\tilde{\theta}_2) = \begin{bmatrix} -c_1 + \tilde{\theta}_2 & 1 & 0 \\ -1 + (c_1 + \tilde{\theta}_2)\tilde{\theta}_2 & -c_2 & 1 \\ 0 & -\gamma & 0 \end{bmatrix}. \quad (7.93)$$

For $c_1 = c_2 = 2$, $\gamma = 6$, and $\theta_2 = 3$, the characteristic polynomials of $A_e(\tilde{\theta}_2)$ is

$$p_e(s) = s^3 + (4 - \tilde{\theta}_2)s^2 + \left((\tilde{\theta}_2)^2 - 7\tilde{\theta}_2 + 11\right)s + 12 - 6\tilde{\theta}_2. \quad (7.94)$$

We calculate $\mathcal{S}^\alpha$, $\alpha = \{s, u, su, c\}$ as follows:

$$\begin{aligned} \mathcal{S}^s &= \left\{(z, \tilde{\theta}_1, \tilde{\theta}_2) \in M \mid \tilde{\theta}_2 \in (-\infty, 2)\right\} \\ \mathcal{S}^{su} &= \left\{(z, \tilde{\theta}_1, \tilde{\theta}_2) \in M \mid \tilde{\theta}_2 \in (2, 6.8791)\right\} \\ \mathcal{S}^u &= \left\{(z, \tilde{\theta}_1, \tilde{\theta}_2) \in M \mid \tilde{\theta}_2 \in (6.8791, +\infty)\right\} \\ \mathcal{S}^c &= \left\{(z, \tilde{\theta}_1, \tilde{\theta}_2) \in M \mid \tilde{\theta}_2 \in \{2, 6.8791\}\right\}. \end{aligned} \quad (7.95)$$

□

Examples 4.1 and 4.2 illustrate, respectively, the case $r = 0$ and $0 < r < p$. A qualitative difference between these two cases while not apparent from the above examples, will be revealed in subsequent sections.

7.6 Invariant Manifolds

While Theorem 7.2 establishes that each solution of the adaptive system converges to an equilibrium point on M, Theorem 7.6 determines whether a given equilibrium point is stable or unstable. We now determine which parts of M are attractive and which are repulsive.

Let us consider an equilibrium point $X^e = (z, \tilde{\theta}_1, \tilde{\theta}_2) = (0, 0, \tilde{\theta}_2^e) \in M \setminus \mathcal{S}^c$. By the center manifold theorem, [37, Theorem 3.2.1], there exist local invariant manifolds $W_{\text{loc}}^s(X^e)$ (stable), $W_{\text{loc}}^u(X^e)$ (unstable), and $W_{\text{loc}}^c(X^e)$ (center). Denote $\zeta = [z^T, \tilde{\theta}_1^T]^T$. By [2, Theorem 2.7.2][1], the flow of the system (7.69) is *topologically equivalent*[2] to the flow of the system

$$\begin{aligned} \dot{\zeta}^s &= -\zeta^s, & \zeta^s &\in W_{\text{loc}}^s(X^e) & (7.96) \\ \dot{\zeta}^u &= \zeta^u, & \zeta^u &\in W_{\text{loc}}^u(X^e) & (7.97) \\ \dot{\tilde{\theta}}_2 &= 0, & \tilde{\theta}_2 &\in W_{\text{loc}}^c(X^e). & (7.98) \end{aligned}$$

While $\dim\{W_{\text{loc}}^c(X^e)\} = p - r$, the dimensions of $W_{\text{loc}}^s(X^e)$ and $W_{\text{loc}}^u(X^e)$ are:

---
[1] For a detailed proof, see [60, Theorem 4.1].
[2] Two flows, $\phi_t(x)$ and $\psi_t(x)$, are said to be topologically (C^0) equivalent if there exists a homeomorphism h, taking orbits of φ_t onto those of ψ_t, preserving their orientation.

7.6 INVARIANT MANIFOLDS

1. If $X^e \in \mathcal{S}^s$, then $\dim\{W^s_{\text{loc}}(X^e)\} = n+r$ and $\dim\{W^u_{\text{loc}}(X^e)\} = 0$.
2. If $X^e \in \mathcal{S}^u$, then $\dim\{W^s_{\text{loc}}(X^e)\} = 0$ and $\dim\{W^u_{\text{loc}}(X^e)\} = n+r$.
3. If $X^e \in \mathcal{S}^{su}$, then $0 < \dim\{W^s_{\text{loc}}(X^e)\}, \dim\{W^u_{\text{loc}}(X^e)\} < n+r$.

Only solutions along stable invariant manifolds can converge to points in $M \setminus \mathcal{S}^c$. These solutions are described by the sets

$$U^s = \bigcup_{t \leq 0} \phi_t\left(W^s_{\text{loc}}(\mathcal{S}^s)\right) \tag{7.99}$$

$$U^u = \bigcup_{t \leq 0} \phi_t\left(W^s_{\text{loc}}(\mathcal{S}^{su})\right) \tag{7.100}$$

where $\phi_t(\cdot)$ is the flow generated by (7.69). The remaining solutions, those converging to $\mathcal{S}^c$, belong to the set denoted by U^c. We point out that $U^s \cup U^u \cup U^c = \mathbb{R}^{n+p}$.

With the next two lemmas we show that the set $\mathcal{S}^c$ has Lebesgue measure zero in $\mathbb{R}^{n+p}$. We prove the lemmas for the case $r = 0$. The extension to the general case is straightforward but more notationally intensive.

Lemma 7.11 *The set $\mathcal{S}^c$ has measure zero in M.*

Proof. It is a tedious exercise to show that the matrix $\dfrac{\partial W(0, \tilde{\theta})^{\text{T}} \tilde{\theta}}{\partial z}$ has the form

$$\frac{\partial W(0, \tilde{\theta})^{\text{T}} \tilde{\theta}}{\partial z} = \{p_{ij}(T_{ij})\}_{n \times n} \tag{7.101}$$

where $p_{ij}(T_{ij})$ is a polynomial function of degree $i - j + 1$ with the argument

$$T_{ij} = \{t_{kl}, \; k = i, \ldots, j, \; l = k, \ldots, j\} . \tag{7.102}$$

In addition, p_{ii} is an identity function, i.e., $p_{ii}(T_{ii}) = t_{ii}$, and $p_{ij} = 0$ for $j > i$. The quantities t_{kl} are defined as

$$t_{kl} = \frac{\partial \varphi_k(0, \ldots, 0)^{\text{T}} \tilde{\theta}}{\partial x_l}, \tag{7.103}$$

so $\dfrac{\partial W(0, \tilde{\theta})^{\text{T}} \tilde{\theta}}{\partial z}$ is a polynomial in $\tilde{\theta}$ that vanishes for $\tilde{\theta} = 0$. For example, in the case $n = 2$, we have

$$\frac{\partial W(0, \tilde{\theta})^{\text{T}} \tilde{\theta}}{\partial z} = \begin{bmatrix} t_{11} & 0 \\ t_{21} + (t_{11} - t_{22})\left(c_1 + \frac{\partial \varphi_1(0)^{\text{T}} \theta}{\partial x_1} - t_{11}\right) & t_{22} \end{bmatrix}. \tag{7.104}$$

With (7.101) and the above properties of $p_{ij}(T_{ij})$, it is easy to verify that

$$\det\left(sI - A_e(\tilde{\theta})\right)^{-1} = s^n + (d_{n-1} + \eta_1(T_{1n}))\, s^{n-1} + \cdots$$
$$+ (d_1 + \eta_{n-1}(T_{1n}))\, s^1 + d_0 + \eta_n(T_{1n}), \tag{7.105}$$

where $\det(sI - A_0)^{-1} = s^n + d_{n-1}s^{n-1} + \cdots + d_1 s^1 + d_0$ (recall (7.56)), and each $\eta_i(T_{1n})$ is a polynomial function of its argument, vanishing at $T_{1n} = 0$. Using Viète's formulae, it is easy to see that if all the coefficients of a polynomial are zero, then all of its roots have zero real part (the converse is not true in the case of conjugate pairs on the imaginary axis). Thus, the set $\mathcal{S}^c$ is contained in the set

$$\mathcal{S}^0 = \left\{ \tilde{\theta} \in \mathbb{R}^p \mid d_i + \eta_{n-i}(T_{1n}(\tilde{\theta})),\ i = 1, \ldots, n \right\}. \tag{7.106}$$

Since $d_i > 0$ and $\eta_{n-i}(T_{1n}(\tilde{\theta}))$ is a *polynomial* function of $\tilde{\theta}$, vanishing at $\tilde{\theta} = 0$, then the set $\mathcal{S}^0$ consists of n hypersurfaces of dimension no higher than $p-1$ in the p-dimensional manifold M. Thus, $\mathcal{S}^c \in \mathcal{S}^0$ has measure zero in M. □

Lemma 7.12 *The set $\mathcal{U}^c$ has measure zero in $\mathbb{R}^{n+p}$.*

Proof. Consider an equilibrium $X^e = (x, \tilde{\theta}) = (0, \tilde{\theta}^e) \in \mathcal{S}^s$. We know that it may have a stable invariant manifold $W^s(X^e)$, an unstable invariant manifold $W^u(X^e)$, and it has a center manifold $W^c(X^e)$ of dimension at least $p+1$. We first show that no solution in $W^c(X^e)$ can converge to X^e. We start by noting that at least one of the eigenvalues of $A_e(\tilde{\theta}^e)$ has a zero real part, and proceed with a proof for the case of a single zero eigenvalue (the extension to other possible cases is straightforward). Denote by $\zeta \in \mathbb{R}^n$ a vector such that

$$A_e(\tilde{\theta}^e)\zeta = A_0 \zeta + \frac{\partial W(0, \tilde{\theta}^e)^\mathrm{T} \tilde{\theta}^e}{\partial z} \zeta = 0. \tag{7.107}$$

Consider now formally

$$\frac{d\left(\tilde{\theta}^\mathrm{T} \Gamma^{-1} \tilde{\theta}\right)}{d\left(z^\mathrm{T} z\right)} = \frac{\frac{1}{2}\frac{d}{dt}\left(\tilde{\theta}^\mathrm{T} \Gamma^{-1} \tilde{\theta}\right)}{\frac{1}{2}\frac{d}{dt}\left(z^\mathrm{T} z\right)} = \frac{-z^\mathrm{T} W(z, \tilde{\theta})^\mathrm{T} \tilde{\theta}}{z^\mathrm{T}\left[A_z(z, \tilde{\theta})z + W(z, \tilde{\theta})^\mathrm{T} \tilde{\theta}\right]}. \tag{7.108}$$

With the mean value theorem we get

$$\frac{d\left(\tilde{\theta}^\mathrm{T} \Gamma^{-1} \tilde{\theta}\right)}{d\left(z^\mathrm{T} z\right)} = \frac{-z^\mathrm{T} \frac{\partial W(0, \tilde{\theta})^\mathrm{T} \tilde{\theta}}{\partial z} z - z^\mathrm{T} G_2(z, \tilde{\theta})}{z^\mathrm{T} A_e(\tilde{\theta}) z + z^\mathrm{T}\left[G_1(z, \tilde{\theta}) + G_2(z, \tilde{\theta})\right]}. \tag{7.109}$$

where G_1 and G_2 are functions such that

$$G_1(0, \tilde{\theta}^e) \equiv G_2(0, \tilde{\theta}^e) \equiv 0 \quad \text{and} \quad \frac{\partial G_1(0, \tilde{\theta}^e)}{\partial z} \equiv \frac{\partial G_2(0, \tilde{\theta}^e)}{\partial z} \equiv 0. \tag{7.110}$$

Let δ be a positive constant. From (7.107) we obtain

$$\left.\frac{d\left(\tilde{\theta}^\mathrm{T} \Gamma^{-1} \tilde{\theta}\right)}{d\left(z^\mathrm{T} z\right)}\right|_{z=\delta\zeta, \tilde{\theta}=\tilde{\theta}^e} = \frac{\delta\zeta^\mathrm{T} A_0 \delta\zeta - \delta\zeta^\mathrm{T} G_2(\delta\zeta, \tilde{\theta}^e)}{\delta\zeta^\mathrm{T}\left[G_1(\delta\zeta, \tilde{\theta}^e) + G_2(\delta\zeta, \tilde{\theta}^e)\right]}. \tag{7.111}$$

7.6 INVARIANT MANIFOLDS

Upon the substitution of (7.56), it follows that

$$\left|\left(\frac{d\left(\tilde{\theta}^T \Gamma^{-1}\tilde{\theta}\right)}{d\left(z^T z\right)}\right|_{z=\delta\zeta,\tilde{\theta}=\tilde{\theta}^e}\right)\right| \geq \frac{c_0 \delta^2 |\zeta|^2 - |\delta\zeta^T G_2(\delta\zeta, \tilde{\theta}^e)|}{|\delta\zeta^T \left[G_1(\delta\zeta, \tilde{\theta}^e) + G_2(\delta\zeta, \tilde{\theta}^e)\right]|}. \quad (7.112)$$

Recalling (7.110), we see that the numerator in (7.112) is quadratic while the denominator is cubic in δ. Therefore

$$\left|\left(\frac{d\left(\tilde{\theta}^T \Gamma^{-1}\tilde{\theta}\right)}{d\left(z^T z\right)}\right|_{z=\delta\zeta,\tilde{\theta}=\tilde{\theta}^e}\right)\right| > 0 \quad \text{for sufficiently small } \delta > 0. \quad (7.113)$$

If a trajectory in $W^c(X^e)$ were to converge to X^e, it would need to do so tangentially to ζ (orthogonal to M, as a consequence of the fact that stable and unstable invariant manifolds of equilibria around X^e are orthogonal to M). In other words, as $\delta \to 0$, the expression in (7.111) would need to tend to zero. This is impossible due to (7.113) which is not only positive but even tends to infinity as $\delta \to 0$! Only solutions in $W^s(X^e)$ can converge to $X^e \in \mathcal{S}^c$, that is, $U^c = W^s(\mathcal{S}^c)$. The manifold $W^s(X^e)$ is of dimension no higher than $n-1$, and since, by Lemma 7.11, $\mathcal{S}^c$ is of measure zero on the p-dimensional manifold M, then by Theorem B.4, $U^c = W^s(\mathcal{S}^c)$ is of measure zero in $\mathbb{R}^{n+p}$. □

Example 7.13 (Example 7.8, cont'd.) We illustrate Lemma 7.12 with Figure 7.3 where solutions in the neighborhood of the critical point $(x, \tilde{\theta}) = (0, 1)$ are given for both $\varphi(x) = x - |x|x$ (solid) and $\varphi(x) = x + |x|x$ (dashed), starting from the same initial conditions. Note that the entire phase plane is the center manifold of the critical equilibrium. While for $\varphi(x) = x + |x|x$ we proved that the critical point is unstable, so some solutions certainly do not converge to it, even the solutions for $\varphi(x) = x - |x|x$, for which the equilibrium is stable, go "around" the critical equilibrium, heading towards equilibria with a stabilizing value of $\tilde{\theta}$. This is confirmed by the fact that

$$\left.\frac{d\left(\tilde{\theta}^2\right)}{d\left(x^2\right)}\right|_{\tilde{\theta}=1} = 8\left(1 \pm \frac{1}{|x|}\right) > 0, \quad |x| < 1. \quad (7.114)$$

No solutions converge to $\mathcal{S}^c = (x, \tilde{\theta}) = (0, 1)$. Thus U^c has measure zero. □

We have partitioned the solutions into the following three sets:

1. U^c: solutions that converge to equilibria in $\mathcal{S}^c$; this set has Lebesgue measure zero by Lemma 7.12.

2. U^u: solutions that converge to unstable equilibria in $\mathcal{S}^{su} \in M$; this set has measure zero in $\mathbb{R}^{n+p}$ because for each $X^e \in \mathcal{S}^{su} \subset M$ (note that $\dim M = p - r$), the set U^s is a manifold of dimension no larger than $n-1$; the conclusion about the measure follows from Theorem B.4.

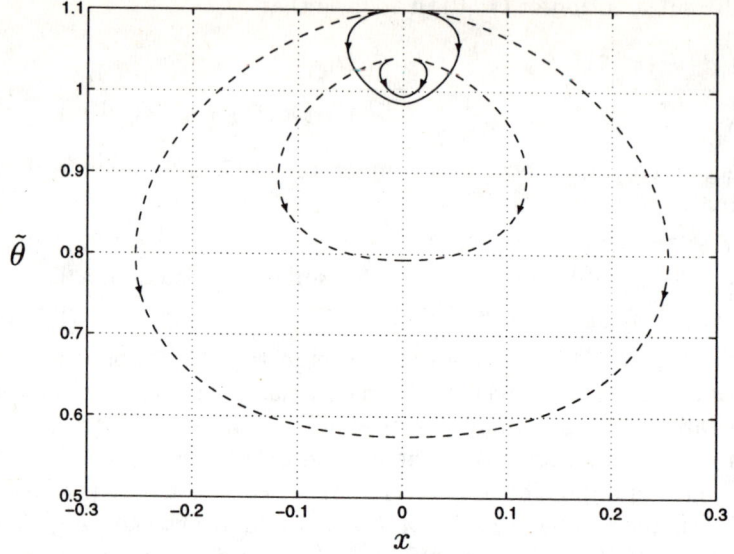

Figure 7.3: Solutions for $\varphi(x) = x - |x|x$ (solid) and $\varphi(x) = x + |x|x$ (dashed). As proved in Lemma 7.12, neither converge to the critical point $(x, \tilde{\theta}) = (0, 1)$.

3. U^s: solutions that converge to stable equilibria in M; this set has a positive Lebesgue measure in $\mathbb{R}^{n+p}$ because, as we showed above, $\mathbb{R}^{n+p} \setminus U^s = U^c \cup U^u$ has measure zero.

We summarize the above argument into the following theorem.

Theorem 7.14 *Consider the adaptive system (7.10)–(7.11). Solutions starting from almost all initial conditions $(z(0), \tilde{\theta}(0)) \in \mathbb{R}^{n+p}$ converge to the set of stable equilibria $\mathcal{S}^s$. The set of initial conditions that generate solutions converging to either $\mathcal{S}^{su}$ or $\mathcal{S}^c$ has Lebesgue measure zero in $\mathbb{R}^{n+p}$. No solutions converge to $\mathcal{S}^u$.*

7.7 Convergence to Stabilizing/Destabilizing Parameters

Now we address the main question of this chapter: Does the adaptive controller 'converge' to a stabilizing nonadaptive (constant) controller?

Let us consider system (7.10) with $\gamma = 0$ (which means $\Gamma = 0$) and $\tilde{\theta}_1 = 0$. Recalling from the definition (7.13) that $\sigma_{jk}(z, \tilde{\theta})$ has Γ as a factor, in view of (7.12), we conclude that $A_z(z, \tilde{\theta})\big|_{\Gamma=0} \equiv A_0$. Therefore, the system (7.44) with

7.7 Convergence to Stabilizing/Destabilizing Parameters

$\gamma = 0$ and $\tilde{\theta}_1 = 0$, becomes

$$\begin{aligned}\dot{z} &= \left(A_z(z,\tilde{\theta}_1,\tilde{\theta}_2)z + W_1^T\tilde{\theta}_1 + W_2^T\tilde{\theta}_2\right)\Big|_{\gamma,\tilde{\theta}_1=0} \\ &= A_0 z + W_2(z,0,\tilde{\theta}_2)^T\tilde{\theta}_2\Big|_{\gamma=0}.\end{aligned} \quad (7.115)$$

The linearization of system (7.115) around the equilibrium $z=0$ is

$$\delta\dot{z} = A_l(\tilde{\theta}_2)\delta z, \quad (7.116)$$

where

$$A_l(\tilde{\theta}_2) = A_0 + \frac{\partial W_2(0,0,\tilde{\theta}_2)^T\tilde{\theta}_2}{\partial z}\Big|_{\gamma=0}. \quad (7.117)$$

Similar to (7.82)–(7.85), we introduce

$$\Lambda^s = \bigcap_{i=1}^n \left\{(z,\tilde{\theta}_1,\tilde{\theta}_2) \in M \mid \text{Re}\,\lambda_i(A_l(\tilde{\theta}_2)) < 0\right\} \quad (7.118)$$

$$\Lambda^u = \bigcap_{i=1}^n \left\{(z,\tilde{\theta}_1,\tilde{\theta}_2) \in M \mid \text{Re}\,\lambda_i(A_l(\tilde{\theta}_2)) > 0\right\} \quad (7.119)$$

$$\Lambda^c = \bigcup_{i=1}^n \left\{(z,\tilde{\theta}_1,\tilde{\theta}_2) \in M \mid \text{Re}\,\lambda_i(A_l(\tilde{\theta}_2)) = 0\right\} \quad (7.120)$$

$$\Lambda^{su} = M \setminus (\Lambda^s \cup \Lambda^u \cup \Lambda^c). \quad (7.121)$$

The values of $\tilde{\theta}_2$ in Λ^s correspond to nonadaptive controllers which are (locally asymptotically) stabilizing. The values of $\tilde{\theta}_2$ in $\Lambda^{su} \cup \Lambda^u$ are destabilizing. Our goal is to see whether all, or almost all, solutions converge to Λ^s. Since Theorem 7.14 establishes that almost all solutions converge to S^s, our task is to determine whether $S^s \subset \Lambda^s$. This translates into the question whether Hurwitzness of $A_e(\tilde{\theta}_2)$ implies Hurwitzness of $A_l(\tilde{\theta}_2)$. We deal with this question in the next three subsections.

7.7.1 Case $r = p$

This case is trivial. By Theorem 4.1, $\hat{\theta}(t) \to \theta$, which is a stabilizing value.

7.7.2 Case $r = 0$

In this case we have the following lemma.

Lemma 7.15 *The linearization of the system* $\dot{z} = A_0 z + W(z,\hat{\theta})^T\Big|_{\Gamma=0}\tilde{\theta}$ *is* $\delta\dot{z} = A_e(\hat{\theta})\delta z$.

Proof. The linearization around the equilibrium $x = 0$ is

$$\delta \dot{z} = \left(A_0 + \frac{\partial W(0, \tilde{\theta})^{\mathrm{T}}\big|_{\Gamma=0} \tilde{\theta}}{\partial z} \right) \delta z. \quad (7.122)$$

Because $F(x)$ is independent of Γ, and $F(0) = 0$, then, in view of (7.9), we have

$$\frac{\partial W(0, \tilde{\theta})^{\mathrm{T}}\big|_{\Gamma=0} \tilde{\theta}}{\partial z} = N(0, \tilde{\theta})\big|_{\Gamma=0} \left(\frac{\partial F(0)^{\mathrm{T}} \tilde{\theta}^e}{\partial x} \right) \frac{\partial x}{\partial z}\bigg|_{\substack{z=0 \\ \Gamma=0}}$$

$$= \begin{bmatrix} 1 & 0 & \cdots & 0 \\ -\frac{\partial \alpha_1}{\partial x_1} & 1 & \ddots & \vdots \\ \vdots & \ddots & \ddots & 0 \\ -\frac{\partial \alpha_{n-1}}{\partial x_1} & \cdots & -\frac{\partial \alpha_{n-1}}{\partial x_{n-1}} & 1 \end{bmatrix}_{\substack{x=0 \\ \Gamma=0}} \left(\frac{\partial F(0)^{\mathrm{T}} \tilde{\theta}^e}{\partial x} \right)$$

$$\times \begin{bmatrix} 1 & 0 & \cdots & 0 \\ \frac{\partial \alpha_1}{\partial z_1} & 1 & \ddots & \vdots \\ \vdots & \ddots & \ddots & 0 \\ \frac{\partial \alpha_{n-1}}{\partial z_1} & \cdots & \frac{\partial \alpha_{n-1}}{\partial z_{n-1}} & 1 \end{bmatrix}_{\substack{z=0 \\ \Gamma=0}}, \quad (7.123)$$

where the last matrix is obtained as an expression for $\frac{\partial x}{\partial z}$ by noting that $x_i = z_i + \alpha_{i-1}$. By carefully examining definitions of the stabilizing functions α_i, one can observe that for $i = 1, \ldots, n$, $k = 1, \ldots, i-1$, and for all $\Gamma \in \mathbb{R}^{p \times p}$,

$$\frac{\partial \alpha_i}{\partial x_k}\bigg|_{\substack{x=0 \\ \Gamma=0}} = \frac{\partial \alpha_i}{\partial x_k}\bigg|_{x=0} \quad (7.124)$$

$$\frac{\partial \alpha_i}{\partial z_k}\bigg|_{\substack{z=0 \\ \Gamma=0}} = \frac{\partial \alpha_i}{\partial z_k}\bigg|_{z=0}. \quad (7.125)$$

By substituting (7.124)–(7.125) into (7.123), we obtain

$$\frac{\partial W(0, \tilde{\theta})^{\mathrm{T}}\big|_{\Gamma=0} \tilde{\theta}}{\partial z} = N(0, \tilde{\theta}) \frac{\partial F(0)^{\mathrm{T}} \tilde{\theta}}{\partial x} \frac{\partial x}{\partial z}\bigg|_{x=0} = \frac{\partial W(0, \tilde{\theta})^{\mathrm{T}} \tilde{\theta}}{\partial z}, \quad (7.126)$$

which, in view of

$$A_e(\tilde{\theta}^e) = A_0 + \frac{\partial W(0, \tilde{\theta}^e)^{\mathrm{T}} \tilde{\theta}^e}{\partial z}. \quad (7.127)$$

and (7.122), completes the proof. □

Therefore the linearization of the frozen system

$$\dot{z} = A_0 z + W(z, \tilde{\theta})^{\mathrm{T}}\big|_{\Gamma=0} \tilde{\theta}. \quad (7.128)$$

7.7 CONVERGENCE TO STABILIZING/DESTABILIZING PARAMETERS

around the equilibrium $x = 0$ is

$$\delta \dot{z} = A_e(\tilde{\theta})\delta z. \tag{7.129}$$

Recalling Theorem 7.6, we conclude that the sufficient conditions for local exponential stability/instability of the equilibrium $x = 0$ of the nonadaptive system are the same as the sufficient conditions for global stability/instability of the equilibrium $(x, \hat{\theta}) = (0, \hat{\theta}^e)$ of the adaptive system.

Let us introduce the sets: $\hat{\Theta}^s$ as the set of values of $\hat{\theta}$ such that all eigenvalues of $A_e(\tilde{\theta})$ have negative real parts; $\hat{\Theta}^u$ as the set of values of $\hat{\theta}$ such that all eigenvalues of $A_e(\tilde{\theta})$ have positive real parts; $\hat{\Theta}^{su}$ as the set of values of $\hat{\theta}$ such that $A_e(\tilde{\theta})$ has eigenvalues with both negative and positive real parts but no zero real parts; $\hat{\Theta}^c$ as the set of values of $\hat{\theta}$ such that $A_e(\tilde{\theta})$ has eigenvalues with zero real parts. These sets can be related to the sets of points on M as follows:

$$\hat{\Theta}^\alpha = \left\{\hat{\theta} \in \mathbb{R}^p \mid (0, \hat{\theta}) \in \mathcal{S}^\alpha\right\}, \quad \alpha \in \{s, u, su, c\}. \tag{7.130}$$

For $\hat{\theta} \in \hat{\Theta}^s$, the nonadaptive controller achieves local exponential stability of $x = 0$. For $\hat{\theta} \in \hat{\Theta}^{su} \cup \hat{\Theta}^u$, the nonadaptive controller results in instability of $x = 0$. For $\hat{\theta} \in \hat{\Theta}^c$, the nonadaptive controller may be either locally stabilizing or destabilizing.

The following corollary now follows from Theorem 7.14.

Corollary 7.16 *Consider the adaptive system (7.10)–(7.11). For almost all initial conditions $(x(0), \hat{\theta}(0)) \in \mathbb{R}^{n+p}$, the parameter estimate $\hat{\theta}(t)$ converges to $\hat{\Theta}^s$. The set of initial conditions such that $\hat{\theta}(t)$ converges to either $\hat{\Theta}^{su}$ or $\hat{\Theta}^c$ has Lebesgue measure zero in $\mathbb{R}^{n+p}$. The parameter estimate $\hat{\theta}(t)$ never converges to $\hat{\Theta}^u$.*

This result is somewhat surprising. Even though the adaptive controller globally stabilizes the equilibrium $(x, \hat{\theta}) = (0, \theta)$, its parameter estimate $\hat{\theta}(t)$ may converge to values in $\hat{\Theta}^{su}$ which result in instability of the equilibrium $x = 0$ of the nonadaptive system! This can happen, however, only along the invariant manifold U^u whose measure is zero in the state space of the adaptive system. Furthermore, since U^u is repulsive, any perturbation will divert the solution away from $\mathcal{S}^{su}$, and $\hat{\theta}(t)$ will converge to $\hat{\Theta}^s$. Still, it is fair to say that the presence of U^u influences the asymptotic behavior of the adaptive system in the sense that, if the initial condition is very close to U^u, then the solution will stay close to it for a long time, but only to be eventually repelled by $\mathcal{S}^{su}$ and attracted to $\mathcal{S}^s$.

It is important to note that, while the nonadaptive controller with $\hat{\theta} = \theta$ guarantees global asymptotic stability (because the closed-loop system is $\dot{z} = A_0 z$), all other $\hat{\theta} \in \hat{\Theta}^s$, in general, guarantee only local asymptotic stability.

Remark 7.17 A special case of particular interest is when the plant is linear, namely, the case where $F(x)^T\theta = \Phi(\theta)x$, and each entry of the matrix $\Phi(\theta)$ is linear in θ. By virtue of (7.9), we have

$$W(z,\tilde{\theta})^T\big|_{\Gamma=0}\tilde{\theta} = N(z,\tilde{\theta})\big|_{\Gamma=0} F(x)^T\tilde{\theta} = N(z,\tilde{\theta})\big|_{\Gamma=0} \Phi(\tilde{\theta})x. \quad (7.131)$$

By examining the definitions of the stabilizing functions α_i, one can observe that for $\Gamma = 0$, they are linear in x, which implies that there exists a nonsingular matrix $T_1(\hat{\theta})$ such that $z = T_1(\hat{\theta})x$, and moreover, there exists a nonsingular matrix $T_2(\hat{\theta})$ such that $N(z,\tilde{\theta})\big|_{\Gamma=0} = T_2(\hat{\theta})$. Therefore

$$W(z,\tilde{\theta})^T\big|_{\Gamma=0}\tilde{\theta} = T_2(\hat{\theta})\Phi(\tilde{\theta})T_1(\hat{\theta})^{-1}z, \quad (7.132)$$

which implies that the closed-loop nonadaptive system (7.128) is linear:

$$\dot{z} = A_0 z + T_2(\hat{\theta})\Phi(\tilde{\theta})T_1(\hat{\theta})^{-1}z \triangleq A_e(\tilde{\theta})z. \quad (7.133)$$

Therefore, the conclusions of Corollary 7.16 are global: for *almost all* initial conditions, the parameter estimate converges to *globally* asymptotically stabilizing values; etc. □

With the following example we illustrate the fact that some solutions converge to $\mathcal{S}^{su}$.

Example 7.18 (Example 7.9, cont'd) Let us return to the adaptive system (7.87) for which we determined that the set $\mathcal{S}^{su}$ is the interval $(-\sqrt{2}, \sqrt{2})$ on the $\hat{\theta}$-axis, which implies that $\hat{\Theta}^{su} = (-\sqrt{2}, \sqrt{2})$. Let us consider a point $(z, \hat{\theta}) = (0, \hat{\theta}^e)$ in $\mathcal{S}^{su}$. With (7.88) we calculate the eigenspaces

$$E^s = \text{span}\left\{\begin{bmatrix} 1 \\ \frac{\hat{\theta}-2-\sqrt{8-3\hat{\theta}^2}}{2} \\ \hat{\theta} \end{bmatrix}\right\}, \quad E^u = \text{span}\left\{\begin{bmatrix} 1 \\ \frac{\hat{\theta}-2+\sqrt{8-3\hat{\theta}^2}}{2} \\ \hat{\theta} \end{bmatrix}\right\} \quad (7.134)$$

of the matrix $\text{diag}\{A_e(\hat{\theta}), 0\}$ corresponding to the eigenvalues λ_1 and λ_2, respectively. Therefore, there exist stable and unstable invariant manifolds, W^s and W^u at each point in $\mathcal{S}^{su}$. These manifolds are shown in Figure 7.4 for a set of points in the interval $(0.6, 1) \subset \hat{\Theta}^{su}$. Even though these points are unstable, solutions along W^s converge to them. The parameter estimate values that they attain are destabilizing for the nonadaptive system. However, as we established in Theorem 7.14, the solutions converging to $\mathcal{S}^{su}$ have measure zero — almost all solutions converge to $\mathcal{S}^s$. This set, defined in (7.90), is the interval $\hat{\Theta}^s = (\sqrt{2}, +\infty)$ on the $\hat{\theta}$-axis, which corresponds to the stabilizing parameter estimate values for the nonadaptive system. In fact, according to Remark 7.17, the parameter estimate values from $\hat{\Theta}^s$ guarantee *global* asymptotic stability because the plant (7.86) with $\varphi_1(x_1) = x_1$ is linear. □

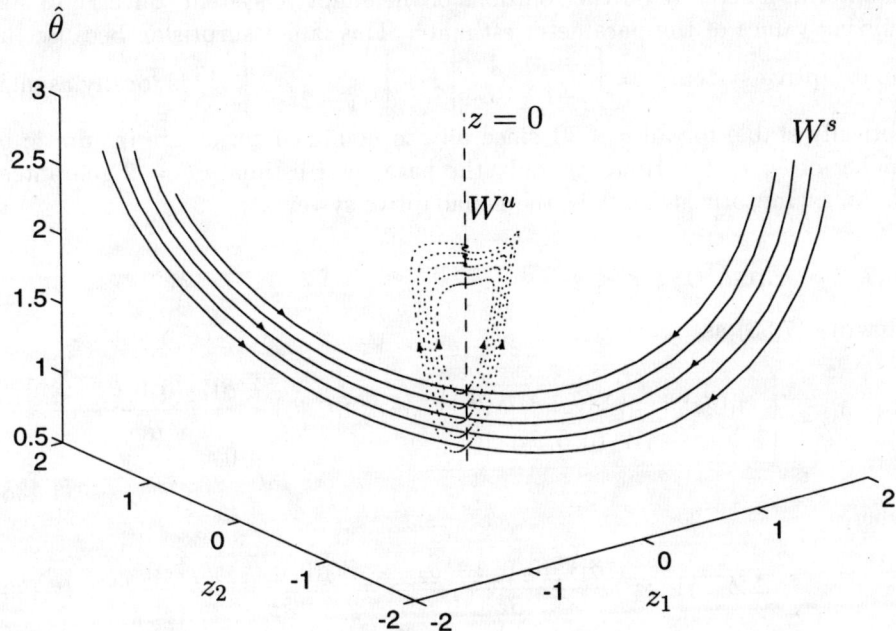

Figure 7.4: Stable (W^s) and unstable (W^u) invariant manifolds of unstable equilibria in $\mathcal{S}^{su}$. Solutions that converge to the unstable equilibria along W^s attain parameter estimate values that are destabilizing for the nonadaptive system. The manifolds W^s and W^u are determined by simulations with initial conditions close to the $\hat{\theta}$-axis, and on the eigenspaces E^s and E^u. Manifold W^s was obtained by simulation in reverse time.

Next we give an example where the entire $\hat{\theta}$-axis is $\mathcal{S}^s$.

Example 7.19 Let us consider the second order system with an unknown scalar parameter:
$$\begin{aligned} \dot{x}_1 &= x_2 + x_1^2 \theta \\ \dot{x}_2 &= u. \end{aligned} \quad (7.135)$$

After the tuning functions design, the resulting error system is:
$$\begin{aligned} \dot{z} &= \begin{bmatrix} -c_1 & 1 \\ -1 & -c_2 \end{bmatrix} z + \begin{bmatrix} 1 \\ -\frac{\partial \alpha_1}{\partial x_1} \end{bmatrix} x_1^2 \tilde{\theta} \\ \dot{\tilde{\theta}} &= -\gamma x_1^2 \left[1, -\frac{\partial \alpha_1}{\partial x_1}\right] z \end{aligned} \quad (7.136)$$

and
$$A_e(\tilde{\theta}) = \begin{bmatrix} -c_1 & 1 \\ -1 & -c_2 \end{bmatrix}. \quad (7.137)$$

Thus $A_e(\tilde{\theta})$ is Hurwitz for all $\tilde{\theta} \in \mathbb{R}$ and all the points on the $\hat{\theta}$-axis are stable equilibria. Therefore all the solutions of the adaptive system converge to stabilizing values of the parameter estimate. This is not surprising because the nonadaptive system $\dot{z} = \begin{bmatrix} -c_1 & 1 \\ -1 & -c_2 \end{bmatrix} z + \begin{bmatrix} 1 \\ c_1 + 2z_1\hat{\theta} \end{bmatrix} z_1^2 \tilde{\theta}$ is locally asymptotically stable for all $\hat{\theta} \in \mathbb{R}$ since all the nonlinear terms are quadratic or higher order in z_1. However, only the parameter estimate $\hat{\theta} = \theta$ guarantees *global* asymptotic stability of the nonadaptive system. □

7.7.3 Case $0 < r < p$

Rewrite (7.55) as

$$A_e(\tilde{\theta}_2) = \begin{bmatrix} A_l(\tilde{\theta}_2) + \gamma\left(\Sigma(\tilde{\theta}_2) + \Delta(\tilde{\theta}_2)\right) & W_1^{\mathrm{T}}(0,0,\tilde{\theta}_2) + \dfrac{\partial W_2(0,0,\tilde{\theta}_2)^{\mathrm{T}} \tilde{\theta}_2}{\partial \tilde{\theta}_1} \\ -\gamma W_1(0,0,\tilde{\theta}_2) & 0 \end{bmatrix}, \tag{7.138}$$

where

$$\gamma\Delta(\tilde{\theta}_2,\gamma) = \frac{\partial W_2(0,0,\tilde{\theta}_2)^{\mathrm{T}} \tilde{\theta}_2}{\partial z} - \left.\frac{\partial W_2(0,0,\tilde{\theta}_2)^{\mathrm{T}} \tilde{\theta}_2}{\partial z}\right|_{\gamma=0}. \tag{7.139}$$

The relationship between the stability properties of $A_e(\tilde{\theta}_2)$ and $A_l(\tilde{\theta}_2)$ is complicated, and no simple conclusions can be drawn. In fact, it is conceivable that, for some $\tilde{\theta}$, $A_e(\tilde{\theta}_2)$ would be Hurwitz while $A_l(\tilde{\theta}_2)$ is not Hurwitz. Let us define the set of equilibria corresponding to this situation:

$$\mathcal{R}^s = \mathcal{S}^s \bigcap \left(\Lambda^u \bigcup \Lambda^{su}\right), \tag{7.140}$$

and denote by $\mathcal{U}^s$ the set of solutions converging to $\mathcal{R}^s$. If $\mathcal{R}^s$ were to have nonzero measure in the $(p-r)$-dimensional manifold M, $\mathcal{U}^s$ would have nonzero measure in $\mathbb{R}^{n+p}$ because the stable invariant manifolds of equilibria in $\mathcal{R}^s$ are $(n+r)$-dimensional. In this case the adaptive controller would be converging to destabilizing nonadaptive controllers from a set of initial conditions of positive measure! The next example illustrates this possibility.

Example 7.20 (Example 7.10, cont'd) Let us return to the adaptive system (7.92) for which we determined the sets $\mathcal{S}^\alpha$ in (7.95). From (7.55) and (7.117) we get

$$A_l(\tilde{\theta}_2) = \begin{bmatrix} -c_1 + \tilde{\theta}_2 & 1 \\ -1 + (c_1 + \hat{\theta}_2)\tilde{\theta}_2 & -c_2 \end{bmatrix}. \tag{7.141}$$

For $c_1 = c_2 = 2$, $\gamma = 6$, and $\theta_2 = 3$, the characteristic polynomial of $A_l(\tilde{\theta}_2)$ is

$$p_l(s) = s^2 + (4 - \tilde{\theta}_2)s + (\tilde{\theta}_2)^2 - 7\tilde{\theta}_2 + 5. \tag{7.142}$$

We calculate Λ^α, $\alpha = \{s, u, su\}$ as follows:

$$\begin{aligned}
\Lambda^s &= \left\{(z, \tilde{\theta}_1, \tilde{\theta}_2) \in M \mid \tilde{\theta}_2 \in (-\infty, 0.8074)\right\} \\
\Lambda^{su} &= \left\{(z, \tilde{\theta}_1, \tilde{\theta}_2) \in M \mid \tilde{\theta}_2 \in (0.8074, 6.1926)\right\} \quad (7.143) \\
\Lambda^u &= \left\{(z, \tilde{\theta}_1, \tilde{\theta}_2) \in M \mid \tilde{\theta}_2 \in (6.1926, +\infty)\right\}.
\end{aligned}$$

In light of (7.95) and (7.143), we have

$$\begin{aligned}
\mathcal{R}^s &= \mathcal{S}^s \cap \left(\Lambda^u \cup \Lambda^{su}\right) \\
&= \left\{(z, \tilde{\theta}_1, \tilde{\theta}_2) \in M \mid \tilde{\theta}_2 \in (0.8074, 2)\right\}. \quad (7.144)
\end{aligned}$$

The set $\mathcal{R}^s$ is an interval $\tilde{\theta}_2 \in (0.8074, 2)$ with positive measure in M. Each point on this interval has a 3-dimensional stable invariant manifold. Therefore, the set $\mathcal{U}^s$ of initial conditions leading to destabilizing estimates has positive measure in $\mathbb{R}^4$.

Another set of interest is

$$\begin{aligned}
\mathcal{R}^{su} &= \mathcal{S}^{su} \cap \Lambda^u \\
&= \left\{(z, \tilde{\theta}_1, \tilde{\theta}_2) \in M \mid \tilde{\theta}_2 \in (6.1926, 6.8791)\right\}. \quad (7.145)
\end{aligned}$$

Along their stable invariant manifolds, the equilibria in $\mathcal{R}^{su}$ attract some solutions denoted by $\mathcal{U}^{su}$. Since $\mathcal{R}^{su} \subset \Lambda^u$, these solutions result in parameter estimates such that *all* of the eigenvalues of the linearized nonadaptive system are unstable. This is different from Section 7.7.2 where no solutions could converge to such "completely destabilizing" parameter estimates. However, $\mathcal{U}^{su}$ has measure zero in $\mathbb{R}^4$. □

This example demonstrates that for $0 < r < p$ the set of initial conditions leading to destabilizing estimates may have positive measure. This is a surprising result considering that the measure of such initial conditions is zero for both $r = 0$ and $r = p$.

7.8 Disconnecting Adaptation

The last problem we address is of practical importance: determine a (sufficiently large) time instant T so that adaptation can be disconnected *at any time after* T without destroying the closed-loop system stability. We study this problem only for the case $r = 0$ because this is the only case where the equilibrium is the same with and without adaptation.

Let us consider the following two systems: the closed-loop *adaptive* system

$$\Sigma : \quad \begin{aligned} \dot{z} &= A_z(z, \hat{\theta})z + W(z, \hat{\theta})^{\mathrm{T}} \tilde{\theta} \\ \dot{\hat{\theta}} &= -\Gamma W(z, \hat{\theta}) z, \end{aligned} \quad (7.146)$$

and the closed-loop *nonadaptive* system

$$\Sigma_{\hat{\theta}}: \qquad \dot{z} = A_0 z + W(z, \hat{\theta})^T \big|_{\Gamma=0} \tilde{\theta}. \qquad (7.147)$$

We observed earlier that system $\Sigma_{\hat{\theta}}$ is obtained by setting $\Gamma = 0$ in system Σ.

Theorem 7.21 *Consider the adaptive system (7.146). For each initial condition* $(x(0), \hat{\theta}(0))$ *in* U^s, *i.e., for* almost all *initial conditions in* $\mathbb{R}^{n+p}$, *there exists a finite time instant* $T \geq 0$ *such that, if* Γ *is set to zero at ANY time instant* $t_1 \geq T$, *the solution* $x(t)$ *of the nonadaptive system (7.147) remains bounded and* $\lim_{t \to \infty} x(t) = 0$.

While it is almost obvious that adaptation can be disconnected at SOME time instant T, here we stress that adaptation can be disconnected at ANY time after some T.

Proof. By Theorem 7.14, all solutions of (7.146) with initial conditions in U^s converge to points in $\mathcal{S}^s$, which is a set of stable equilibria. Let us consider a solution $X(t) = (x(t), \hat{\theta}(t))$ converging to $X_\infty = (0, \hat{\theta}_\infty) \in \mathcal{S}^s$. The idea of the proof is presented pictorially in Figure 7.5.

Since X_∞ is a stable equilibrium, and the set $\mathcal{S}^s$ is open in M, then there exists $R > 0$ such that

$$|X(t_0) - X_\infty| < R \;\; \Rightarrow \;\; |X(t) - X_\infty| < \frac{1}{2}\mathrm{dist}\{X_\infty, \partial \mathcal{S}^s\}, \quad \forall t \geq t_0, \qquad (7.148)$$

where $\partial \mathcal{S}^s$ denotes the boundary of $\mathcal{S}^s$. Since $\mathrm{dist}\{X_\infty, \partial \mathcal{S}^s\} = \mathrm{dist}\{\hat{\theta}_\infty, \partial \hat{\Theta}^s\}$, then (7.148) means that, once the solution $X(t)$ enters the ball of radius R around X_∞, the parameter estimate $\hat{\theta}(t)$ will remain inside the set $\hat{\Theta}^s$ of (locally) stabilizing values. Because the equilibrium $x = 0$ for $\Sigma_{\hat{\theta}}$, $\hat{\theta} \in \hat{\Theta}^s$ is, in general, only *locally* stable, we cannot say that T is the instant when $X(t)$ enters the ball of radius R around X_∞, because $x(T)$ may be outside the region of attraction for $\Sigma_{\hat{\theta}}$.

Let us define the set

$$\hat{\Theta}_\Pi = \left\{ \hat{\theta} \in \hat{\Theta}^s \;\middle|\; |\hat{\theta} - \hat{\theta}_\infty| < \frac{1}{2}\mathrm{dist}\{\hat{\theta}_\infty, \partial \hat{\Theta}^s\} \right\}, \qquad (7.149)$$

which is the set where $\hat{\theta}(t)$ remains after $X(t)$ enters the ball of radius R around X_∞. Denote by $\rho_{\hat{\theta}}$ the radius of the largest ball (around the origin) which is strictly inside the region of attraction of the equilibrium $x = 0$ for the nonadaptive system $\Sigma_{\hat{\theta}}$. Since $\hat{\Theta}_\Pi$ is chosen as in (7.149), i.e., separated from $\hat{\Theta}^s$ by a boundary layer, then

$$\rho \stackrel{\triangle}{=} \inf_{\hat{\theta} \in \hat{\Theta}_\Pi} \rho_{\hat{\theta}} \qquad (7.150)$$

7.8 DISCONNECTING ADAPTATION

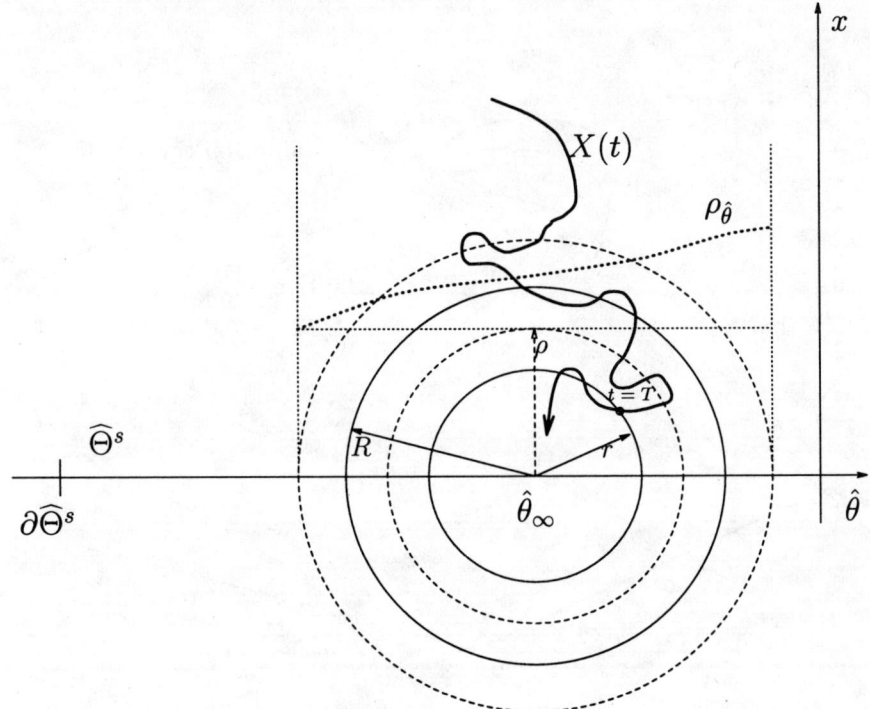

Figure 7.5: Adaptation can be disconnected after $X(t)$ has entered the ball where both $\hat{\theta}$ is an asymptotically stabilizing parameter estimate for $\Sigma_{\hat{\theta}}$, and x is inside the region of attraction of $x = 0$ for $\Sigma_{\hat{\theta}}$.

is positive. By virtue of the stability of the equilibrium X_∞, there exists $r > 0$ such that

$$|X(t_0) - X_\infty| < r \quad \Rightarrow \quad |X(t) - X_\infty| < \rho, \quad \forall t \geq t_0. \tag{7.151}$$

From (7.148) and (7.151) it follows that

$$|X(T) - X_\infty| < \min\{R, r\} \quad \Rightarrow \quad |X(t_1) - X_\infty| < \frac{1}{2}\min\left\{\operatorname{dist}\left\{\hat{\theta}_\infty, \partial\widehat{\Theta}^s\right\}, \rho\right\}$$
$$\Rightarrow \quad \hat{\theta}(t_1) \in \widehat{\Theta}_{\Pi} \quad \text{and} \quad |x(t_1)| < \rho, \tag{7.152}$$

for all $t_1 \geq T$. Hence, for any $t_1 \geq T$, $\hat{\theta}$ will be an asymptotically stabilizing parameter estimate for $\Sigma_{\hat{\theta}}$, and x will be inside the region of attraction of $x = 0$. Since $\min\{R, r\} > 0$ and $X(t) - X_\infty \to 0$, then $T < \infty$, which completes the proof. $\square$

Let us now illustrate Theorem 7.21 with an example.

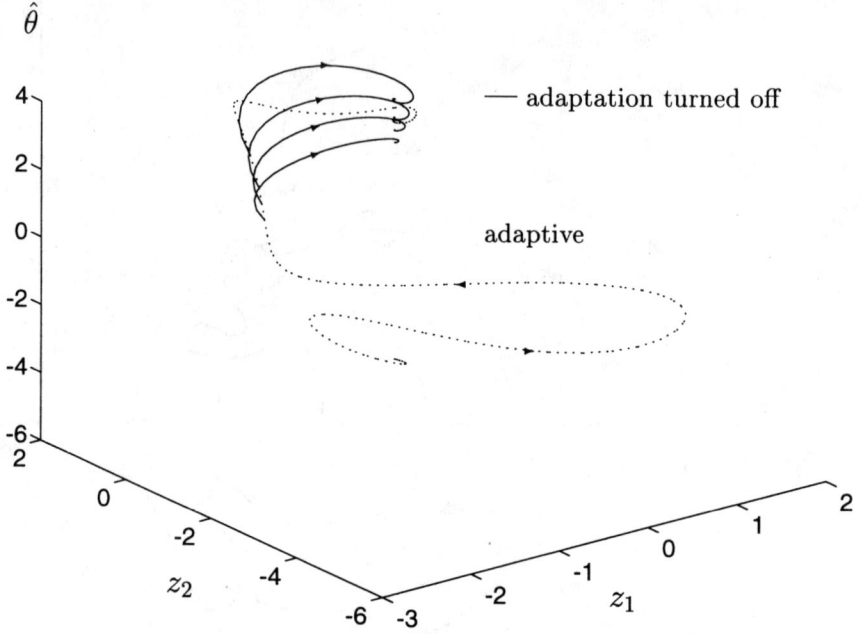

Figure 7.6: Behavior after disconnecting adaptation. The dotted curve shows a trajectory of the adaptive system. The disconnection of adaptation freezes the parameter estimate at four different stabilizing values, and the four nonadaptive trajectories (solid curves) remain bounded and converge to $z = 0$.

Example 7.22 (Example 7.9, cont'd) Let us return to the adaptive system (7.87) for which we determined that the set $\mathcal{S}^s$ is the interval $\left(\sqrt{2}, \infty\right)$ on the $\hat{\theta}$-axis, which implies that $\widehat{\Theta}^s = \left(\sqrt{2}, \infty\right)$. The dotted curve in Figure 7.6 shows a trajectory of the adaptive system. This trajectory starts from $(z_1, z_2, \hat{\theta}) = (.1, .1, -4.5)$ and converges to $(z_1, z_2, \hat{\theta}) = (0, 0, 2.5657) \in \mathcal{S}^s$. The four solid curves represent trajectories of the nonadaptive system after the adaptation has been turned off. The disconnection of adaptation after $T = 3.7$ freezes the parameter estimate at four different values in $\widehat{\Theta}^s$, and, as expected, the trajectories converge to $z = 0$ (in four different horizontal planes). It is very important to note that in this example we have dealt with a linear plant $(\varphi_1(x_1) = x_1)$, so according to Remark 7.17, the parameter estimate values in $\widehat{\Theta}^s$ guarantee *global* asymptotic stability of the nonadaptive system. □

Notes and References

There exists virtually no previous literature (other than the work of the authors) on the subject of this chapter. The chapter is based on Krstić [62] ($r=0$) and Li and Krstić [71] ($0 < r < p$). For related work on nonlinear dynamical systems issues in adaptive systems with unmodeled effects, the reader is referred to the reference list in the book of Mareels and Polderman [75]. Conditions for the convergence of the parameter estimates to their true values in adaptive backstepping schemes can be found in Lin and Kanellakopoulos [72].

Chapter 8

Extremum Seeking

The mainstream methods of adaptive control for linear [5, 35, 42] and nonlinear [63] systems (see also Chapter 5 in this book) are applicable only for regulation to *known* set points or reference trajectories. In some applications, the reference-to-output map has an *extremum* (w.l.o.g. we assume that it is a maximum) and the objective is to select the set point to keep the output at the extremum value. The uncertainty in the reference-to-output map makes it necessary to use some sort of adaptation to find the set point which maximizes the output.

This is a very different optimal control problem than the one we considered in Chapter 5 where we were dealing with stabilization (and tracking) for the class of systems

$$\dot{x} = F(x)\theta + g(x)u \qquad (8.1)$$

where the objective was to optimize the *transient* energy expended on global stabilization (both the energy of the state x and the energy of control u). In this chapter we deal with more general classes of systems (non-affine) but the control objective is only steady-state optimization (achieving the extremum value for a given system output with respect to some controller parameter) and the stability objective is local.

In this chapter we present an extremum seeking scheme, provide elementary design guidelines, and study its stability via averaging and singular perturbation techniques.

8.1 Extremum Seeking—Problem Statement

Consider a general SISO nonlinear model

$$\dot{x} = f(x, u) \qquad (8.2)$$
$$y = h(x), \qquad (8.3)$$

where $x \in \mathbb{R}^n$ is the state, $u \in \mathbb{R}$ is the input, $y \in \mathbb{R}$ is the output, and $f : \mathbb{R}^n \times \mathbb{R} \to \mathbb{R}^n$ and $h : \mathbb{R}^n \to \mathbb{R}$ are smooth. Suppose that we know a smooth control law

$$u = \alpha(x, \theta) \tag{8.4}$$

parameterized by a parameter θ. To reduce notation, we assume in this chapter that θ is scalar. (The extension to the vector case is trivial by making a in Figure 8.1 a vector of the same dimension as θ.) The closed-loop system

$$\dot{x} = f(x, \alpha(x, \theta)) \tag{8.5}$$

then has equilibria parameterized by θ. We make the following assumptions about the closed-loop system.

Assumption 8.1 *There exists a smooth function $l : \mathbb{R} \to \mathbb{R}^n$ such that*

$$f(x, \alpha(x, \theta)) = 0 \quad \text{if and only if} \quad x = l(\theta). \tag{8.6}$$

Assumption 8.2 *For each $\theta \in \mathbb{R}$, the equilibrium $x = l(\theta)$ of the system (8.5) is locally exponentially stable.*

Hence, we assume that we have a control law (8.4) which is robust with respect to its own parameter θ in the sense that it *exponentially stabilizes any of the equilibria that θ may produce*. Except for the requirement that Assumption 8.2 holds *for any* $\theta \in \mathbb{R}$ (which we impose only for notational convenience and can easily relax to an *interval* in $\mathbb{R}$), this assumption is not restrictive. It simply means that we have a control law designed for local stabilization and this control law need not be based on modeling knowledge of either $f(x, u)$ or $l(\theta)$. State feedback (8.4) can also be replaced by dynamic output feedback without loss of generality.

The next assumption is central to the problem of extremum seeking.

Assumption 8.3 *There exists $\theta^* \in \mathbb{R}$ such that*

$$(h \circ l)'(\theta^*) = 0 \tag{8.7}$$
$$(h \circ l)''(\theta^*) < 0. \tag{8.8}$$

Thus, we assume that the output equilibrium map $y = h(l(\theta))$ has a *maximum* at $\theta = \theta^*$. The objective is to develop a feedback mechanism which maximizes the steady state value of y but without requiring the knowledge of either θ^* or the functions h and l. The assumption that $h \circ l$ has a maximum is without loss of generality—the case with a minimum would be treated identically by replacing y by $-y$ in the subsequent feedback design.

8.2 A Peak Seeking Scheme

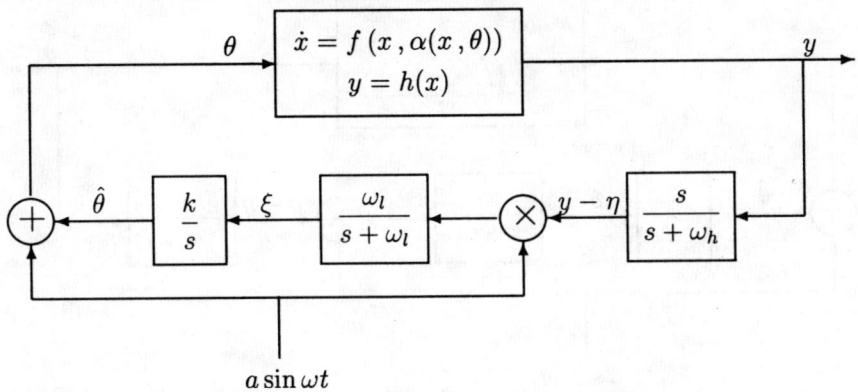

Figure 8.1: A peak seeking feedback scheme.

8.2 A Peak Seeking Scheme

The peak seeking feedback scheme is shown in Figure 8.1. Before we engage in extensive efforts to prove stability of the scheme, we explain its basic idea. We start by pointing out that it is impossible to conclude that a certain point is a maximum without visiting the neighborhood on both sides of the maximum. For this reason, we employ a *slow* periodic perturbation $a \sin \omega t$ which is added to the signal $\hat{\theta}$, our best estimate of θ^* (the persistent nature of $a \sin \omega t$ may be undesirable but is necessary to maintain a maximum in the face of changes in functions f and h). If the perturbation is slow, then the plant appears as a static map $y = h \circ l(\theta)$ (see Figure 8.2) and its dynamics do not interfere with the peak seeking scheme. The peak seeking scheme then makes ξ approximately proportional to $(h \circ l)^*(\hat{\theta})$, as explained next.

Let us start by approximating y by the first order Taylor expansion $h \circ l(\hat{\theta}) + (h \circ l)^*(\hat{\theta}) a \sin \omega t$. The high-pass filter $\dfrac{s}{s+\omega_h}$ eliminates the "DC component" $h \circ l(\hat{\theta})$ off y. Thus, $a \sin \omega t \times \dfrac{s}{s+\omega_h} y$ will be (approximately) equal to $(h \circ l)^*(\hat{\theta}) \dfrac{a^2}{2}[1 + \cos(2\omega t)]$. Then the "DC component" is extracted by the low-pass filter $\dfrac{\omega_l}{s+\omega_l}$ so that ξ is approximately equal to $\dfrac{a^2}{2}(h \circ l)^*(\hat{\theta})$. Thus, $\hat{\theta}$ is tuned in the direction of maximizing y.

Despite its apparent simplicity, the proof of "stability" of this feedback scheme is nontrivial even for the static case in Figure 8.2. As we shall see in the sequel, both the analysis of the scheme and the selection of design parameters are indeed intricate. These parameters are selected as

$$\omega_h = \omega \omega_H = \omega \delta \omega'_H = O(\omega \delta) \tag{8.9}$$

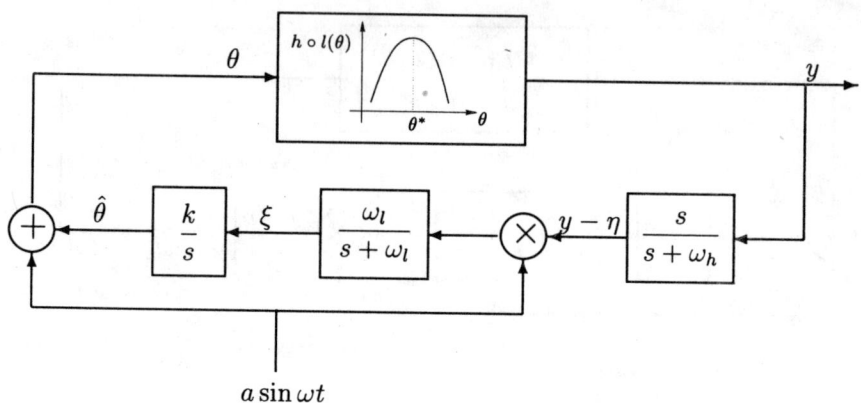

Figure 8.2: If perturbation $a \sin \omega t$ is slow, the plant can be viewed as a static map.

$$\omega_l = \omega \omega_L = \omega \delta \omega'_L = O(\omega \delta) \quad (8.10)$$
$$k = \omega K = \omega \delta K' = O(\omega \delta), \quad (8.11)$$

where ω and δ are small positive constants and ω'_H, ω'_L, and K' are $O(1)$ positive constants. As it will become apparent later, a also needs to be small.

From (8.9) and (8.10) we see that the cut-off frequencies of the filters need to be lower than the frequency of the perturbation signal. In addition, the adaptation gain k needs to be small. Thus, the overall feedback system has three time scales:

- fastest—the plant with the stabilizing controller,
- medium—the periodic perturbation,
- slow—the filters in the peak seeking scheme.

The analysis that follows treats first the static case from Figure 8.2 using the method of *averaging* (Section 4). Then we use the *singular perturbation* method (Section 5) for the full system in Figure 8.1.

Before we start the analysis, we summarize the system in Figure 8.1 as

$$\dot{x} = f\left(x, \alpha(x, \hat{\theta} + a \sin \omega t)\right) \quad (8.12)$$
$$\dot{\hat{\theta}} = k\xi \quad (8.13)$$
$$\dot{\xi} = -\omega_l \xi + \omega_l (y - \eta) a \sin \omega t \quad (8.14)$$
$$\dot{\eta} = -\omega_h \eta + \omega_h y. \quad (8.15)$$

Let us introduce the new coordinates

$$\tilde{\theta} = \hat{\theta} - \theta^* \quad (8.16)$$
$$\tilde{\eta} = \eta - h \circ l(\theta^*). \quad (8.17)$$

8.3 Averaging Analysis

Then, in the time scale $\tau = \omega t$, the system (8.12)–(8.15) is rewritten as

$$\omega \frac{dx}{d\tau} = f\left(x, \alpha(x, \theta^* + \tilde{\theta} + a\sin\tau)\right) \tag{8.18}$$

$$\frac{d}{d\tau}\begin{bmatrix}\tilde{\theta}\\\xi\\\tilde{\eta}\end{bmatrix} = \delta\begin{bmatrix}K'\xi\\-\omega'_L\xi + \omega'_L\left(h(x) - h\circ l(\theta^*) - \tilde{\eta}\right)a\sin\tau\\-\omega'_H\tilde{\eta} + \omega'_H\left(h(x) - h\circ l(\theta^*)\right)\end{bmatrix}. \tag{8.19}$$

8.3 Averaging Analysis

The first step in the analysis is to study the system in Figure 8.2. We "freeze" x in (8.18) at its "equilibrium" value

$$x = l\left(\theta^* + \tilde{\theta} + a\sin\tau\right) \tag{8.20}$$

and substitute it into (8.19), getting the "reduced system"

$$\frac{d}{d\tau}\begin{bmatrix}\tilde{\theta}_r\\\xi_r\\\tilde{\eta}_r\end{bmatrix} = \delta\begin{bmatrix}K'\xi_r\\-\omega'_L\xi_r + \omega'_L\left(\nu\left(\tilde{\theta}_r + a\sin\tau\right) - \tilde{\eta}_r\right)a\sin\tau\\-\omega'_H\tilde{\eta}_r + \omega'_H\nu\left(\tilde{\theta}_r + a\sin\tau\right)\end{bmatrix}, \tag{8.21}$$

where

$$\nu\left(\tilde{\theta}_r + a\sin\tau\right) = h\circ l\left(\theta^* + \tilde{\theta}_r + a\sin\tau\right) - h\circ l(\theta^*). \tag{8.22}$$

In view of Assumption 8.3, it is obvious that

$$\nu(0) = 0 \tag{8.23}$$
$$\nu'(0) = (h\circ l)'(\theta^*) = 0 \tag{8.24}$$
$$\nu''(0) = (h\circ l)''(\theta^*) < 0. \tag{8.25}$$

The system (8.21) is in the form to which the averaging method is applicable. The average model of (8.21) is

$$\frac{d}{d\tau}\begin{bmatrix}\tilde{\theta}_r^a\\\xi_r^a\\\tilde{\eta}_r^a\end{bmatrix} = \delta\begin{bmatrix}K'\xi_r^a\\-\omega'_L\xi_r^a + \frac{\omega'_L}{2\pi}a\int_0^{2\pi}\nu\left(\tilde{\theta}_r^a + a\sin\sigma\right)\sin\sigma\,d\sigma\\-\omega'_H\tilde{\eta}_r^a + \frac{\omega'_H}{2\pi}\int_0^{2\pi}\nu\left(\tilde{\theta}_r^a + a\sin\sigma\right)d\sigma\end{bmatrix}. \tag{8.26}$$

First we need to determine the average equilibrium $\left(\tilde{\theta}_r^{a,e}, \xi_r^{a,e}, \tilde{\eta}_r^{a,e}\right)$ which satisfies

$$\xi_r^{a,e} = 0 \tag{8.27}$$
$$\int_0^{2\pi}\nu\left(\tilde{\theta}_r^{a,e} + a\sin\sigma\right)\sin\sigma\,d\sigma = 0 \tag{8.28}$$
$$\tilde{\eta}_r^{a,e} = \frac{1}{2\pi}\int_0^{2\pi}\nu\left(\tilde{\theta}_r^{a,e} + a\sin\sigma\right)d\sigma. \tag{8.29}$$

By postulating $\tilde{\theta}_r^{a,e}$ in the form

$$\tilde{\theta}_r^{a,e} = b_1 a + b_2 a^2 + O(a^3), \quad (8.30)$$

substituting in (8.28), using (8.23) and (8.24), integrating, and equating the like powers of a, we get $\nu''(0)b_1 = 0$ and $\nu''(0)b_2 + \frac{1}{8}\nu'''(0) = 0$, which implies that

$$\tilde{\theta}_r^{a,e} = -\frac{\nu'''(0)}{8\nu''(0)} a^2 + O(a^3). \quad (8.31)$$

Another round of lengthy calculations applied to (8.29) yields

$$\tilde{\eta}_r^{a,e} = \frac{\nu''(0)}{4} a^2 + O(a^3). \quad (8.32)$$

Thus, the equilibrium of the average model (8.26) is

$$\begin{bmatrix} \tilde{\theta}_r^{a,e} \\ \xi_r^{a,e} \\ \tilde{\eta}_r^{a,e} \end{bmatrix} = \begin{bmatrix} -\frac{\nu'''(0)}{8\nu''(0)} a^2 + O(a^3) \\ 0 \\ \frac{\nu''(0)}{4} a^2 + O(a^3) \end{bmatrix}. \quad (8.33)$$

The Jacobian of (8.26) at $\left(\tilde{\theta}, \xi, \tilde{\eta}\right)_r^{a,e}$ is

$$J_r^a = \delta \begin{bmatrix} 0 & K' & 0 \\ \frac{\omega_L'}{2\pi} a \int_0^{2\pi} \nu'\left(\tilde{\theta}_r^{a,e} + a\sin\sigma\right) \sin\sigma d\sigma & -\omega_L' & 0 \\ \frac{\omega_H'}{2\pi} \int_0^{2\pi} \nu'\left(\tilde{\theta}_r^{a,e} + a\sin\sigma\right) d\sigma & 0 & -\omega_H' \end{bmatrix}. \quad (8.34)$$

Since J_r^a is block-lower-triangular we easily see that it will be Hurwitz if and only if

$$\int_0^{2\pi} \nu'\left(\tilde{\theta}_r^{a,e} + a\sin\sigma\right) \sin\sigma d\sigma < 0. \quad (8.35)$$

More calculations that use (8.23) and (8.24) give

$$\int_0^{2\pi} \nu'\left(\tilde{\theta}_r^{a,e} + a\sin\sigma\right) \sin\sigma d\sigma = \pi\nu''(0)a + O(a^2). \quad (8.36)$$

By substituting (8.36) into (8.34) we get

$$\det(\lambda I - J_r^a) = \left(\lambda^2 + \delta\omega_L'\lambda - \frac{\delta^2 \omega_L' K'}{2}\nu''(0)a^2 + O(\delta^2 a^3)\right)(\lambda + \delta\omega_H'), \quad (8.37)$$

which, in view of (8.25), proves that J_r^a is Hurwitz for sufficiently small a. This, in turn, implies that the equilibrium (8.33) of the average system (8.26) is exponentially stable for a sufficiently small a. Then, according to the Averaging Theorem [58, Theorem 8.3] we have the following result.

8.3 AVERAGING ANALYSIS

Theorem 8.4 *Consider the system (8.21) under Assumption 8.3. There exist $\bar{\delta}$ and $\bar{a}$ such that for all $\delta \in (0, \bar{\delta})$ and $a \in (0, \bar{a})$ the system (8.21) has a unique exponentially stable periodic solution $\left(\tilde{\theta}_r^{2\pi}(\tau), \xi_r^{2\pi}(\tau), \tilde{\eta}_r^{2\pi}(\tau)\right)$ of period 2π and this solution satisfies*

$$\left\| \begin{bmatrix} \tilde{\theta}_r^{2\pi}(\tau) + \frac{\nu'''(0)}{8\nu''(0)} a^2 \\ \xi_r^{2\pi}(\tau) \\ \tilde{\eta}_r^{2\pi}(\tau) - \frac{\nu''(0)}{4} a^2 \end{bmatrix} \right\| \leq O(\delta) + O(a^3), \qquad \forall \tau \geq 0. \tag{8.38}$$

This result implies that all solutions $\left(\tilde{\theta}_r(\tau), \xi_r(\tau), \tilde{\eta}_r(\tau)\right)$, and, in particular, their $\tilde{\theta}_r(\tau)$-components, converge to an $O(\delta + a^2)$-neighborhood of the origin. It is important to interpret this result in terms of the system in Figure 8.2. Since $y = h \circ l \left(\theta^* + \tilde{\theta}_r(\tau) + a \sin \tau \right)$ and $(h \circ l)'(\theta^*) = 0$, we have

$$y - h \circ l(\theta^*) = (h \circ l)''(\theta^*) \left(\tilde{\theta}_r + a \sin \tau \right)^2 + O \left(\left(\tilde{\theta}_r + a \sin \tau \right)^3 \right), \tag{8.39}$$

where

$$\tilde{\theta}_r + a \sin \tau = \left(\tilde{\theta}_r - \tilde{\theta}_r^{2\pi} \right) + \left(\tilde{\theta}_r^{2\pi} + \frac{(h \circ l)'''(\theta^*)}{8(h \circ l)''(\theta^*)} a^2 \right) - \frac{(h \circ l)'''(\theta^*)}{8(h \circ l)''(\theta^*)} a^2 + a \sin \tau. \tag{8.40}$$

Since the first term converges to zero, the second term is $O(\delta + a^3)$, the third term is $O(a^2)$ and the fourth term is $O(a)$, then

$$\limsup_{\tau \to \infty} \left| \tilde{\theta}_r(\tau) + a \sin \tau \right| = O(a + \delta). \tag{8.41}$$

Thus, (8.39) yields

$$\limsup_{\tau \to \infty} |y(\tau) - h \circ l(\theta^*)| = O\left(a^2 + \delta^2\right). \tag{8.42}$$

The last expression characterizes the asymptotic performance of the peak seeking scheme in Figure 8.2 and explains why it is not only important that the periodic perturbation be small but also that the cut-off frequencies of the filters and the adaptation gain k be low.

Another important conclusion can be drawn from (8.38). The solution $\tilde{\theta}_r(\tau)$ will converge $O(\delta + a^3)$-close to $-\frac{(h \circ l)'''(\theta^*)}{8(h \circ l)''(\theta^*)} a^2$. Since $(h \circ l)''(\theta^*) < 0$, the sign of this quantity depends on the sign of $(h \circ l)'''(\theta^*)$. If $(h \circ l)'''(\theta^*) > 0$ (respectively, < 0), then the curve $h \circ l(\theta)$ will be more "flat" on the right (respectively, left) side of $\theta = \theta^*$. Since $\tilde{\theta}_r$ will have an offset in the direction of $\mathrm{sgn}\left\{(h \circ l)'''(\theta^*)\right\}$, then $\tilde{\theta}_r(t)$ will converge to the "flatter" side of $h \circ l(\theta)$. This is precisely what we want—to be on the side where $h \circ l(\theta)$ is less sensitive to variations in θ and closer to its maximum value.

8.4 Singular Perturbation Analysis

Now we address the full system in Figure 8.1 whose state space model is given by (8.18) and (8.19) in the time scale $\tau = \omega t$. To make the notation in our further analysis compact, we write (8.19) as

$$\frac{dz}{d\tau} = \delta G(\tau, x, z), \qquad (8.43)$$

where $z = (\tilde{\theta}, \xi, \tilde{\eta})$. By Theorem 8.4, there exists an exponentially stable periodic solution $z_r^{2\pi}(\tau)$ such that

$$\frac{dz_r^{2\pi}(\tau)}{d\tau} = \delta G\left(\tau, L\left(\tau, z_r^{2\pi}(\tau)\right), z_r^{2\pi}(\tau)\right), \qquad (8.44)$$

where $L(\tau, z) = l\left(\theta^* + \tilde{\theta} + a\sin\tau\right)$. To bring the system (8.18) and (8.43) into the *standard singular perturbation form*, we shift the state z using the transformation

$$\tilde{z} = z - z_r^{2\pi}(\tau) \qquad (8.45)$$

and get

$$\frac{d\tilde{z}}{d\tau} = \delta \tilde{G}(\tau, x, \tilde{z}) \qquad (8.46)$$

$$\omega \frac{dx}{d\tau} = \tilde{F}(\tau, x, \tilde{z}), \qquad (8.47)$$

where

$$\tilde{G}(\tau, x, \tilde{z}) = G\left(\tau, x, \tilde{z} + z_r^{2\pi}(\tau)\right) - G\left(\tau, L\left(\tau, z_r^{2\pi}(\tau)\right), z_r^{2\pi}(\tau)\right) \qquad (8.48)$$

$$\tilde{F}(\tau, x, \tilde{z}) = f\left(x, \alpha\left(x, \theta^* + \underbrace{\tilde{\theta} - \tilde{\theta}_r^{2\pi}(\tau)}_{\tilde{z}_1} + \tilde{\theta}_r^{2\pi}(\tau) + a\sin\tau\right)\right). \qquad (8.49)$$

We note that

$$x = L\left(\tau, \tilde{z} + z_r^{2\pi}(\tau)\right) \qquad (8.50)$$

is the *quasi-steady state*, and that the *reduced model*

$$\frac{d\tilde{z}_r}{d\tau} = \delta \tilde{G}\left(\tau, L\left(\tau, \tilde{z}_r + z_r^{2\pi}(\tau)\right), \tilde{z}_r + z_r^{2\pi}(\tau)\right) \qquad (8.51)$$

has an equilibrium at the origin $\tilde{z}_r = 0$ (cf. (8.48) with (8.50)). This equilibrium has been shown in Section 8.3 to be exponentially stable for sufficiently small a.

To complete the singular perturbation analysis, we also study the *boundary layer model* (in the time scale $t = \tau/\omega$):

$$\frac{dx_b}{dt} = \tilde{F}\left(\tau, x_b + L\left(\tau, \tilde{z} + z_r^{2\pi}(\tau)\right), \tilde{z}\right)$$
$$= f(x_b + l(\theta), \alpha(x_b + l(\theta), \theta)), \qquad (8.52)$$

8.4 SINGULAR PERTURBATION ANALYSIS

where $\theta = \theta^* + \tilde{\theta} + a\sin\tau$ should be viewed as a parameter independent from the time variable t. Since $f(l(\theta), \alpha(l(\theta), \theta)) \equiv 0$, then $x_b = 0$ is an equilibrium of (8.52). By Assumption 8.2, this equilibrium is exponentially stable.

By combining exponential stability of the reduced model (8.51) with the exponential stability of the boundary layer model (8.52), using the Tikhonov type theorem on the infinite interval [58, Theorem 9.4], we conclude the following:

- The solution $z(\tau)$ of (8.43) is $O(\omega)$-close to the solution $z_r(\tau)$ of (8.51), and therefore, it exponentially converges to an $O(\omega)$-neighborhood of the periodic solution $z_r^{2\pi}(\tau)$, which is $O(\delta)$-close to the equilibrium $z_r^{a,e}$. This, in turn, implies that the solution $\tilde{\theta}(\tau)$ of (8.19) exponentially converges to an $O(\omega+\delta)$-neighborhood of $-\dfrac{(h \circ l)'''(\theta^*)}{8(h \circ l)''(\theta^*)}a^2 + O(a^3)$. It follows then that $\theta(\tau) = \theta^* + \tilde{\theta}(\tau) + a\sin\tau$ exponentially converges to an $O(\omega+\delta+a)$-neighborhood of θ^*.

- The solution $x(\tau)$ of (8.47) (which is the same as (8.18)) satisfies

$$x(\tau) - l\left(\theta^* + \tilde{\theta}_r(\tau) + a\sin\tau\right) - x_b(t) = O(\omega), \quad (8.53)$$

where $\tilde{\theta}_r(\tau)$ is the solution of the reduced model (8.21) and $x_b(t)$ is the solution of the boundary layer model (8.52). From (8.53) we get

$$x(\tau) - l(\theta^*) = O(\omega) + l\left(\theta^* + \tilde{\theta}_r(\tau) + a\sin\omega\tau\right) - l(\theta^*) - x_b(t). \quad (8.54)$$

Since $\tilde{\theta}_r(\tau)$ exponentially converges to the periodic solution $\tilde{\theta}_r^{2\pi}(\tau)$, which is $O(\delta)$-close to the average equilibrium $\dfrac{(h \circ l)'''(\theta^*)}{8(h \circ l)''(\theta^*)}a^2 + O(a^3)$, and since the solution $x_b(t)$ of (8.52) is exponentially decaying, then by (8.54), $x(\tau) - l(\theta^*)$ exponentially converges to an $O(\omega+\delta+a)$-neighborhood of zero. Consequently, $y = h(x)$ exponentially converges to an $O(\omega+\delta+a)$-neighborhood of its maximal equilibrium value $h \circ l(\theta^*)$.

We summarize the above conclusions in the following theorem.

Theorem 8.5 *Consider the feedback system (8.12)–(8.15) under Assumptions 8.1–8.3. There exists a ball of initial conditions around the point $\left(x, \hat{\theta}, \xi, \eta\right) = (l(\theta^*), \theta^*, 0, h \circ l(\theta^*))$ and constants $\bar{\omega}, \bar{\delta}$, and $\bar{a}$ such that for all $\omega \in (0, \bar{\omega}), \delta \in (0, \bar{\delta})$, and $a \in (0, \bar{a})$, the solution $\left(x(t), \hat{\theta}(t), \xi(t), \eta(t)\right)$ exponentially converges to an $O(\omega + \delta + a)$-neighborhood of that point. Furthermore, $y(t)$ converges to an $O(\omega + \delta + a)$-neighborhood of $h \circ l(\theta^*)$.*

A considerably more elaborate analysis leads to the following stronger result which we give without proof.

Theorem 8.6 *Under the conditions of Theorem 8.5, there exists a unique exponentially stable periodic solution of (8.12)–(8.15) in an $O(\omega + \delta + a)$-neighborhood of the point $\left(x, \hat{\theta}, \xi, \eta\right) = (l(\theta^*), \theta^*, 0, h \circ l(\theta^*))$.*

Notes and References

The extremum seeking problem, also known as "extremum control" or "self-optimizing control", was popular in the 1950's and 1960's [9, 21, 33, 47, 56, 82, 87, 90], much before the theoretical breakthroughs in adaptive linear control of the 1980's. In fact, the emergence of extremum control dates as far back as the 1922 paper of Leblanc [69], whose scheme may very well have been the first "adaptive" controller reported in the literature. Among the surveys on extremum control, we find the one by Sternby [103] particularly useful, as well as Section 13.3 in Astrom and Wittenmark [5] which puts extremum control among the most promising future areas for adaptive control. Among the many applications of extremum control overviewed in [103] and [5] are combustion processes (for IC engines, steam generating plants, and gas furnaces), grinding processes, solar cell and radio telescope antenna adjustment to maximize the received signal, and blade adjustment in water turbines and wind mills to maximize the generated power. A more recent application of extremum control are anti-lock braking systems where schemes different from the one in this chapter are currently in use [20]. On the theoretical side, the pioneering averaging studies of Meerkov [79, 80, 81] are a precursor to the stability results presented here.

Most of the results available on extremum control consider a plant as a static map. A few references approach problems where the plant is a cascade of a nonlinear static map and a linear dynamic system (the so-called Hammerstein and Wiener models), see [114] and references therein. In this chapter we approach the general problem where the nonlinearity with an extremum is a reference-to-output *equilibrium* map for a general nonlinear (non-affine in control) system stabilizable around each of these equilibria by a local feedback controller.

This chapter is based on a paper by Krstić and Wang [66] which also presents an application of extremum seeking to aeroengine compressors.

Appendices

Appendix A

Legendre-Fenchel Transform and Young's Inequality

For a class $\mathcal{K}_\infty$ function γ whose derivative exists and is also a class $\mathcal{K}_\infty$ function, $\ell\gamma$ denotes the *Legendre-Fenchel transform*

$$\ell\gamma(r) = r(\gamma')^{-1}(r) - \gamma\left((\gamma')^{-1}(r)\right), \tag{A.1}$$

where $(\gamma')^{-1}(r)$ stands for the inverse function of $\dfrac{d\gamma(r)}{dr}$.

Lemma A.1 *If γ and its derivative γ' are class $\mathcal{K}_\infty$ functions, then the Legendre-Fenchel transform satisfies the following properties:*

(a) $\quad \ell\gamma(r) = r(\gamma')^{-1}(r) - \gamma\left((\gamma')^{-1}(r)\right) = \displaystyle\int_0^r (\gamma')^{-1}(s)\,ds \quad$ (A.2)

(b) $\quad \ell\ell\gamma = \gamma \quad$ (A.3)

(c) $\quad \ell\gamma$ *is a class $\mathcal{K}_\infty$ function* (A.4)

(d) $\quad \ell\gamma(\gamma'(r)) = r\gamma'(r) - \gamma(r).$ (A.5)

Proof. (a) Integrating by parts, we get

$$\begin{aligned}\int_0^r (\gamma')^{-1}(s)\,ds &= r(\gamma')^{-1}(r) - \int_0^r s\,d\left((\gamma')^{-1}(s)\right)\\ &= r(\gamma')^{-1}(r) - \int_0^r \gamma'\left((\gamma')^{-1}(s)\right)d\left((\gamma')^{-1}(s)\right)\\ &= r(\gamma')^{-1}(r) - \int_0^r d\left(\gamma\left((\gamma')^{-1}(s)\right)\right), \end{aligned} \tag{A.6}$$

which completes the proof.

(b) Immediate by differentiating the second expression in (a), and then inverting and integrating the result.

(c) Obvious from the second expression in (a) because $(\gamma')^{-1}$ is a class $\mathcal{K}_\infty$ function.

(d) Follows by direct substitution into (A.2). $\square$

Lemma A.2 (Young's inequality [38, Theorem 156]) *For any two vectors x and y, the following holds*

$$x^T y \leq \gamma(|x|) + \ell\gamma(|y|), \qquad (A.7)$$

and the equality is achieved if and only if

$$y = \gamma'(|x|)\frac{x}{|x|}, \text{ that is, for } x = (\gamma')^{-1}(|y|)\frac{y}{|y|}. \qquad (A.8)$$

Corollary A.3 *For any two vectors x and y, the following holds*

$$x^T y \leq \frac{\epsilon^p}{p}|x|^p + \frac{1}{q\epsilon^q}|y|^q, \qquad (A.9)$$

where $\epsilon > 0$ and the constants $p > 1$ and $q > 1$ satisfy $(p-1)(q-1) = 1$.

Appendix B

Measure of Global Invariant Manifolds

Consider the dynamical system

$$\dot{x} = f(x), \qquad x \in \mathbb{R}^n, \tag{A.1}$$

where the vector field f is C^r ($r \geq 1$) on $\mathbb{R}^n$. Let $\phi_t^f : \mathbb{R}^n \to \mathbb{R}^n$ denote the flow of f. For $\bar{x}$ such that $f(\bar{x}) = 0$, let E^s, E^u and E^c denote the (generalized) eigenspaces of $Df(\bar{x})$ corresponding to eigenvalues with negative, positive and zero real parts, respectively. Let s, u, and c denote the dimensions of the spaces E^s, E^u and E^c. Let $\Pi^\alpha(U)$, where U is a neighborhood of $\bar{x}$, denote the projection of U to E^α, $\alpha \in \{s, u, c\}$. The following theorem is well known:

Theorem B.1 ([2, 37, 57, 60]) *There exist (locally) stable, unstable, and center invariant manifolds, $W^s_{\text{loc}}(\bar{x})$, $W^u_{\text{loc}}(\bar{x})$, and $W^c_{\text{loc}}(\bar{x})$, tangent respectively to E^s, E^u and E^c at $\bar{x}$. The manifolds $W^s_{\text{loc}}(\bar{x})$, $W^u_{\text{loc}}(\bar{x})$, and $W^c_{\text{loc}}(\bar{x})$ are graphs of functions $h^s : \Pi^s(U) \to E^u \times E^c$, $h^u : \Pi^u(U) \to E^s \times E^c$, $h^c : \Pi^c(U) \to E^s \times E^u$ such that $h^\alpha(0) = 0$, $Dh^\alpha(0) = 0$, $\alpha = s, u, c$. While $W^s_{\text{loc}}(\bar{x})$ and $W^u_{\text{loc}}(\bar{x})$ are C^r and unique, $W^c_{\text{loc}}(\bar{x})$ is C^{r-1} and may not be unique.*

One would now think of defining (as, e.g., in [37, Eq. (1.3.7)]) the *global invariant manifolds* of $\bar{x}$ by using points on the local manifolds as initial conditions:

$$W^\alpha(\bar{x}) = \bigcup_{t \in T^\alpha} \phi_t^f \left(W^\alpha_{\text{loc}}(\bar{x}) \right) \tag{A.2}$$

where $T^s = \mathbb{R}_-$, $T^u = \mathbb{R}_+$, and $T^c = \mathbb{R}$. However, the flow ϕ_t^f may not be defined for all $t \in \mathbb{R}$, so we cannot use (A.2). Let us therefore consider the system

$$\dot{x} = \frac{1}{1 + f(x)^\mathrm{T} f(x)} f(x) \triangleq F(x), \tag{A.3}$$

where $F(x)$ is C^r and bounded, so its flow $\phi_t^F : \mathbb{R}^n \to \mathbb{R}^n$ is a global diffeomorphism for each $t \in \mathbb{R}$. Since $\frac{1}{1+f(x)^T f(x)}$ is a positive scalar function, the orbits of the modified system (A.3) coincide with the orbits of the system (A.1). (The orientation of the orbits is also preserved.) Therefore, the invariant sets of the systems (A.1) and (A.3) are the same. Hence, $W^s_{\text{loc}}(\bar{x})$, $W^u_{\text{loc}}(\bar{x})$, and $W^c_{\text{loc}}(\bar{x})$ are also local invariant manifolds of (A.3) at $\bar{x}$. Let us now use the flow ϕ_t^F to define the global invariant manifolds of (A.3) as

$$W^s(\bar{x}) = \bigcup_{t \leq 0} \phi_t^F \left(W^s_{\text{loc}}(\bar{x}) \right) \tag{A.4}$$

$$W^u(\bar{x}) = \bigcup_{t \geq 0} \phi_t^F \left(W^u_{\text{loc}}(\bar{x}) \right) \tag{A.5}$$

$$W^c(\bar{x}) = \bigcup_{\forall t} \phi_t^F \left(W^c_{\text{loc}}(\bar{x}) \right) . \tag{A.6}$$

Since the local manifolds $W^\alpha_{\text{loc}}(\bar{x})$ are graphs of functions $h^\alpha(\cdot)$, they are *immersed* manifolds of dimension α in $\mathbb{R}^n$. Since the global manifolds are obtained by propagating the local manifolds with the diffeomorphism ϕ_t^F, the dimension of the global manifold $W^\alpha(\bar{x})$ is α.

Theorem B.2 *There exist stable, unstable, and center global invariant manifolds, $W^s(\bar{x})$, $W^u(\bar{x})$, and $W^c(\bar{x})$, tangent to E^s, E^u and E^c at $\bar{x}$, and of dimension s, u, and c.*

We now characterize the measure of the global invariant manifolds. Consider first the case where not all of the eigenvalues of $Df(\bar{x})$ have negative real part, that is, $s < n$. Since $W^s_{\text{loc}}(\bar{x})$ is the graph of the function $h^s : \Pi^s(U) \to E^u \times E^c$, it has Lebesgue measure zero in $\mathbb{R}^n$:

$$\lambda \{W^s_{\text{loc}}(\bar{x})\} = 0 . \tag{A.7}$$

Let $\mathcal{T} = \{t_i \mid t_i < 0, i = 1, 2, \ldots\}$ be a countable set of points such that $W^s(\bar{x})$ is represented as

$$W^s(\bar{x}) = \bigcup_{t_i \in \mathcal{T}} \phi_{t_i}^F \left(W^s_{\text{loc}}(\bar{x}) \right) . \tag{A.8}$$

The measure of $W^s(\bar{x})$ satisfies

$$\lambda \{W^s(\bar{x})\} \leq \sum_{t_i \in \mathcal{T}} \lambda \left\{ \phi_{t_i}^F \left(W^s_{\text{loc}}(\bar{x}) \right) \right\} . \tag{A.9}$$

Since $\phi_{t_i}^F$ is a diffeomorphism for all t_i, then $\lambda \left\{ \phi_{t_i}^F \left(W^s_{\text{loc}}(\bar{x}) \right) \right\} = 0$. Recalling that $\mathcal{T}$ is countable, we conclude that $\lambda \{W^s(\bar{x})\} = 0$. The same argument applies for $W^u(\bar{x})$ and $W^c(\bar{x})$. Thus we have the following theorem.

Theorem B.3 *Suppose $\alpha < n$, $\alpha \in \{s, u, c\}$. Then $W^\alpha(\bar{x})$ has Lebesgue measure zero in $\mathbb{R}^n$.*

The center manifold $W^c(\bar{x})$ deserves a special attention because of its possible nonuniqueness (cf. Theorem B.1). A classical example is the system [57]

$$\dot{x} = -x, \qquad \dot{y} = y^2, \tag{A.10}$$

where $W^s(0) = \{y = 0\}$ and $W^c(0)$ can be any of the curves

$$x = \begin{cases} k\exp(1/y) & y < 0 \\ 0 & y \geq 0 \end{cases} \tag{A.11}$$

The $y < 0$ half-plane is "filled" with a family of center manifolds parametrized by k. While each of the center manifolds has zero measure, the family has a positive measure.

The following theorem is an immediate corrolary of Theorem B.3.

Theorem B.4 *Consider a set $\mathcal{E}$ of equilibria belonging to a q-dimesional equilibrium manifold $\mathcal{M}$ and consider the corresponding set of global invariant manifolds $W^\alpha(\mathcal{E})$ for some $\alpha \in \{s, u, c\}$. If $\alpha + q < n$ then $W^\alpha(\mathcal{E})$ has Lebesgue measure zero in $\mathbb{R}^n$.*

Bibliography

[1] B. D. O. Anderson and J. B. Moore, *Optimal Control: Linear Quadratic Methods*, Prentice-Hall, Englewood Cliffs, NJ, 1990.

[2] D. K. Arrowsmith and C. M. Place, *An Introduction to Dynamical Systems*, Cambridge University Press, 1990.

[3] Z. Artstein, "Stabilization with relaxed controls," *Nonlinear Analysis*, TMA-7, pp. 1163–1173, 1983.

[4] K. J. Astrom, *Introduction to Stochastic Control Theory*, New York: Academic Press, 1970.

[5] K. J. Astrom and B. Wittenmark, *Adaptive Control*, 2nd edition, Reading, MA: Addison-Wesley, 1995.

[6] J. A. Ball, J. W. Helton, and M. L. Walker, "$\mathcal{H}_\infty$ control for nonlinear systems with output feedback," *IEEE Transactions on Automatic Control*, vol. 38, pp. 546–559, 1993.

[7] T. Başar and P. Bernhard, *$\mathcal{H}_\infty$-Optimal Control and Related Minimax Design Problems*, Boston: Birkhauser, 1995.

[8] D. S. Bernstein, "Robust static and dynamic output-feedback stabilization: deterministic and stochastic perspectives," *IEEE Transactions on Automatic Control*, vol. 32, pp. 1076–1084, 1987.

[9] P. F. Blackman, "Extremum-Seeking Regulators," in J. H. Westcott, Ed., *An Exposition of Adaptive Control*, New York, NY: The Macmillan Company, 1962.

[10] C. I. Byrnes, F. Delli Priscoli and A. Isidori, *Output Regulation of Uncertain Nonlinear Systems*, Boston: Birkhauser, 1997

[11] C. I. Byrnes, A. Isidori, and J. C. Willems, "Passivity, feedback equivalence, and the global stabilization of minimum phase nonlinear systems," *IEEE Transactions on Automatic Control*, vol. 36, pp. 1228–1240, 1991.

[12] J. Carr, *Applications of Centre Manifold Theory*, New York: Springer-Verlag, 1981.

[13] H. Deng and M. Krstić, "Stochastic nonlinear stabilization—Part I: A backstepping design," *Systems & Control Letters*, to appear, 1997.

[14] H. Deng and M. Krstić, "Stochastic nonlinear stabilization—Part II: Inverse optimality," *Systems & Control Letters*, to appear, 1997.

[15] H. Deng and M. Krstić, "Output-feedback stochastic nonlinear stabilization," submitted to *IEEE Transactions on Automatic Control*, 1997, also in *Proceedings of IEEE CDC*, San Diego, CA, 1997.

[16] H. Deng and M. Krstić, "Stabilization of stochastic nonlinear systems driven by noise of unknown covariance," submitted to *IEEE Transactions on Automatic Control* and ACC'98, 1997.

[17] G. Didinsky and T. Başar, "Minimax adaptive control of uncertain plants," *Proceedings of the 33rd IEEE Conference on Decision and Control*, Lake Buena Vista, FL, pp. 2839–2844, December, 1994.

[18] J. L. Doob, *Stochastic Processes*, New York: John Wiley & Sons, 1953.

[19] D. Down, S. P. Meyn and R. L. Tweedie, "Exponential and uniform ergodicity of Markov processes", *The Annals of Probability*, vol. 23, pp. 1671–1691, 1995.

[20] S. Drakunov, U. Ozguner, P. Dix, and B. Ashrafi, "ABS control using optimum search via sliding modes," *IEEE Transactions on Control Systems Technology*, vol. 3, pp. 79–85, 1995.

[21] C. S. Drapper and Y. T. Li, "Principles of optimalizing control systems and an application to the internal combustion engine," *ASME*, vol. 160, pp. 1–16, 1951, also in R. Oldenburger, Ed., *Optimal and Self-Optimizing Control*, Boston, MA: The M.I.T. Press, 1966.

[22] T. E. Duncan and B. Pasik-Duncan, "Stochastic adaptive control," in W. S. Levine (Ed.), *The Control Handbook*, CRC Press, pp. 1127–1136, 1996.

[23] L. El Ghaoui, "State-feedback control of systems with multiplicative noise via linear matrix inequalities," *Systems & Control Letters*, vol. 24, pp. 223–228, 1995.

[24] K. Ezal, Z. Pan and P. V. Kokotovic, "Locally optimal backstepping design," *Proceedings of the 36th IEEE Conference on Decision and Control*, San Diego, CA, pp. 1767–1773, 1997.

BIBLIOGRAPHY

[25] W. H. Fleming and W. M. McEneaney, "Risk-sensitive control on an infinite time horizon," *SIAM Journal on Control and Optimization*, vol. 33, No. 6, pp.1881–1915, 1995.

[26] W. H. Fleming and H. M. Soner, *Control Markov Processes and Viscosity Solutions*, Springer-Verlag, 1993.

[27] P. Florchinger, "Lyapunov-like techniques for stochastic stability," *SIAM Journal of Control and Optimization*, vol. 33, pp. 1151–1169, 1995.

[28] P. Florchinger, "Global stabilization of cascade stochastic systems," *Proceedings of the 34th Conference on Decision & Control*, New Orleans, LA, pp. 2185–2186, 1995.

[29] P. Florchinger, "A universal formula for the stabilization of control stochastic differential equations," *Stochastic Analysis and Applications*, vol. 11, pp. 155–162, 1993.

[30] P. Florchinger, "Feedback stabilization of affine in the control stochastic differential systems by the control Lyapunov function method," *SIAM Journal on Control and Optimization*, vol. 35, pp. 500–511, 1997.

[31] R. A. Freeman and P. V. Kokotovic, "Inverse optimality in robust stabilization," *SIAM Journal on Control and Optimization*, vol. 34, pp. 1365–1391, 1996.

[32] R. A. Freeman and P. V. Kokotović, *Robust Nonlinear Control Design*, Boston, MA: Birkhäuser, 1996.

[33] A. L. Frey, W. B. Deem, and R. J. Altpeter, "Stability and optimal gain in extremum-seeking adaptive control of a gas furnace," *Proceedings of the Third IFAC World Congress*, London, 48A, 1966.

[34] S. T. Glad, "On the gain margin of nonlinear and optimal regulators," *IEEE Transactions on Automatic Control*, vol. 29, pp. 615–620, 1984.

[35] G. C. Goodwin and K. S. Sin, *Adaptive Filtering Prediction and Control*, Englewood Cliffs, NJ: Prentice-Hall, 1984.

[36] M. Green and D. J. N. Limebeer, *Linear Robust Control*, Englewood Cliffs, NJ: Prentice Hall, 1995.

[37] J. Guckenheimer and P. Holmes, *Nonlinear Oscillations, Dynamical Systems, and Bifurcations of Vector Fields*, New York: Springer-Verlag, 1983.

[38] G. Hardy, J. E. Littlewood, and G. Polya, *Inequalities*, 2nd edition, Cambridge University Press, 1989.

[39] U. G. Haussmann and W. Suo, "Singular optimal stochastic controls. I. Existence," *SIAM Journal on Control and Optimization*, vol. 33, pp. 916–936, 1995.

[40] U. G. Haussmann and W. Suo, "Singular optimal stochastic controls. II. Dynamic programming," *SIAM Journal on Control and Optimization*, vol. 33, pp. 937–959, 1995.

[41] J. W. Helton and W. Zhan, "An inequality governing nonlinear $\mathcal{H}_\infty$ control systems," *Systems & Control Letter*, vol. 22, pp. 157–165, 1994.

[42] P. A. Ioannou and J. Sun, *Stable and Robust Adaptive Control*, Englewood Cliffs, NJ: Prentice-Hall, 1995.

[43] A. Isidori, "A necessary condition for nonlinear $\mathcal{H}_\infty$ control via measurement feedback," *Systems & Control Letters*, vol. 23, pp. 169–178, 1994.

[44] A. Isidori, *Nonlinear Control Systems*, 3rd edition, New York: Springer-Verlag, 1995.

[45] A. Isidori and A. Astolfi, "Disturbance attenuation and $\mathcal{H}_\infty$-control via measurement feedback in nonlinear systems," *IEEE Transactions on Automatic Control*, vol. 37, pp. 1283–1293, 1992.

[46] A. Isidori and W. Kang, "$\mathcal{H}_\infty$ control via measurement feedback for general nonlinear systems," *IEEE Transactions on Automatic Control*, vol. 40, pp. 466–472, 1995.

[47] O. L. R. Jacobs and G. C. Shering, "Design of a single-input sinusoidal-perturbation extremum-control system," *Proceedings IEE*, vol. 115, pp. 212-217, 1968.

[48] M. R. James and J. S. Baras, "Robust $\mathcal{H}_\infty$ output feedback control for nonlinear systems," *IEEE Transactions on Automatic Control*, vol. 40, pp. 1007–1017, 1995.

[49] M. R. James and J. S. Baras, "Partial observed differential games, infinite-dimensional Hamilton-Jacobi-Isaac equations, and nonlinear $\mathcal{H}_\infty$ control," *SIAM Journal on Control and Optimization*, vol. 34, pp. 1342–1364, 1996.

[50] M. R. James, J. Baras and R. J. Elliott, "Risk-sensitive control and dynamic games for partially observed discrete-time nonlinear systems," *IEEE Transactions on Automatic Control*, vol. 39, pp. 780–792, 1994.

[51] M. Janković, R. Sepulchre, and P. Kokotović, "CLF based designs with robustness to dynamic input uncertainties," submitted to *Systems and Control Letters*, August 1997.

[52] Z. P. Jiang and I. M. Y. Mareels, "A small-gain control method for nonlinear cascaded systems with dynamic uncertainties," *IEEE Transactions on Automatic Control*, vol. 42, no. 3, pp. 292–308, 1997.

[53] Z. P. Jiang and J.-B. Pomet, "A note on 'Robust control of nonlinear systems with input unmodeled dynamics," *INRIA* Report 2293, June 1994.

[54] Z. P. Jiang, A. R. Teel and L. Praly, "Small-gain theorem for ISS systems and applications," *Mathematics of Control, Signals, and Systems*, vol. 7, pp. 95–120, 1995.

[55] I. Kanellakopoulos, P. V. Kokotović, and A. S. Morse, "Systematic design of adaptive controllers for feedback linearizable systems," *IEEE Transactions on Automatic Control*, vol. 36, pp. 1241–1253, 1991.

[56] V. V. Kazakevich, "Extremum control of objects with inertia and of unstable objects," *Soviet Physics*, Dokl. 5, pp. 658-661, 1960.

[57] A. Kelley, "The stable, center-stable, center, center-unstable, and unstable manifolds," an appendix to *Transversal Mappings and Flows* by R. Abraham and J. Robbin, New York: Benjamin, 1967.

[58] H. K. Khalil, *Nonlinear Systems*, 2nd edition, Englewood Cliffs, NJ: Prentice Hall, 1995.

[59] R. Z. Khas'minskii, *Stochastic Stability of Differential Equations*, Rockville, Maryland: S & N International publisher, 1980.

[60] U. Kirchgraber and K. J. Palmer, *Geometry in the Neighborhood of Invariant Manifolds of Maps and Flows and Linearization*, Longman, 1990.

[61] A. J. Krener, "Necessary and sufficient conditions for nonlinear worst case ($\mathcal{H}_\infty$) control and estimation," *J. Math. Systems Estim. Control*, vol. 4, no. 4, pp. 485–488, 1994.

[62] M. Krstić, "Invariant manifolds and asymptotic properties of adaptive nonlinear systems," *IEEE Transactions on Automatic Control*, vol. 41, pp. 817–829, 1996.

[63] M. Krstić, I. Kanellakopoulos, and P. V. Kokotović, *Nonlinear and Adaptive Control Design*, New York: Wiley, 1995.

[64] M. Krstić and Z. H. Li, "Inverse optimal design of input-to-state stabilizing nonlinear controllers," to appear in *IEEE Transactions on Automatic Control*, 1998, also in *Proc. 1997 IEEE CDC*.

[65] M. Krstić, J. Sun, and P. V. Kokotović, "Robust control of nonlinear systems with input unmodeled dynamics," *IEEE Transactions on Automatic Control*, vol. 41, pp. 913–920, 1996.

[66] M. Krstić and H. H. Wang, "Extremum seeking feedback: stability proof and application to an aeroengine compressor model," submitted to *Automatica*, 1997 (also 1997 CDC).

[67] H. J. Kushner, *Stochastic Stability and Control*, New York: Academic, 1967.

[68] J. P. LaSalle, "Stability theory for ordinary differential equations", *Journal of Differential Equations*, vol. 4, pp. 57–65, 1968.

[69] M. Leblanc, "Sur l'electrification des chemins de fer au moyen de courants alternatifs de frequence elevee," *Revue Generale de l'Electricite*, 1922.

[70] Z. H. Li and M. Krstić, "Optimal design of adaptive tracking controllers for nonlinear systems," *Automatica*, vol. 33, no. 8, pp. 1459–1473, 1997.

[71] Z. H. Li and M. Krstić, "Geometric/asymptotic properties of adaptive nonlinear systems with partial excitation," to appear in *IEEE Trans. Automatic Control*, 1998; also *Proc. 35th IEEE CDC*, pp. 4683–4688, 1996.

[72] J. S. Lin and I. Kanellakopoulos, "Nonlinearities enhance parameter convergence in strict-feedback systems," to appear in *IEEE Trans. Automatic Control*, 1998.

[73] W. M. Lu and J. C. Doyle, "$\mathcal{H}_\infty$ control of nonlinear systems: a convex characterization," *IEEE Transactions on Automatic Control*, vol. 40, pp. 1668–1675, 1995.

[74] X. Mao, *Stability of Stochastic Differential Equations with Respect to Semimartingales*, Longman, 1991.

[75] I. Mareels and J. W. Polderman, *Adaptive Systems: An Introduction*, Boston: Birkhauser, 1996

[76] R. Marino, W. Respondek, A. J. van der Schaft, and P. Tomei, "Nonlinear $\mathcal{H}_\infty$ almost disturbance decoupling," *Systems & Control Letters*, vol. 23, pp. 159–168, 1994.

[77] R. Marino and P. Tomei, *Nonlinear Control Design: Geometric, Adaptive, and Robust*, Prentice Hall, London, 1995.

[78] F. Mazenc and L. Praly, "Adding integrations, saturated controls, and stabilization for feedforward systems," *IEEE Transactions on Automatic Control*, vol. 41, pp. 1559–1578, 1996.

[79] S. M. Meerkov, "Asymptotic methods for investigating quasistationary states in continuous systems of automatic optimization," *Automation and Remote Control*, no. 11, pp. 1726-1743, 1967.

[80] S. M. Meerkov, "Asymptotic methods for investigating a class of forced states in extremal systems," *Automation and Remote Control*, no. 12, pp. 1916–1920, 1967.

[81] S. M. Meerkov, "Asymptotic methods for investigating stability of continuous systems of automatic optimization subjected to disturbance action," (in Russian) *Avtomatika i Telemekhanika*, no. 12, pp. 14–24, 1968.

[82] I. S. Morosanov, "Method of extremum control," *Automation & Remote Control*, vol. 18, pp. 1077–1092, 1957.

[83] P. J. Moylan and B. D. O. Anderson, "Nonlinear regulator theory and an inverse optimal control problem," *IEEE Transactions on Automatic Control*, vol. 18, pp. 460–465, 1973.

[84] H. Nagai, "Bellman equations of risk-sensitive control," *SIAM Journal of Control and Optimization*, vol. 34, pp. 74–101, 1996.

[85] H. Nijmeijer and A. van der Schaft, *Nonlinear Dynamical Control Systems*, New York: Springer-Verlag, 1990.

[86] B. Øksendal, *Stochastic Differential Equations–An Introduction with Applications*, New York: Springer-Verlag, 1995.

[87] I. I. Ostrovskii, "Extremum regulation," *Automation & Remote Control*, vol. 18, pp. 900–907, 1957.

[88] Z. Pan and T. Başar, "Adaptive controller design for tracking and disturbance attenuation in parametric-strict-feedback nonlinear systems," *Proceedings of the 13th IFAC Congress of Automatic Control*, San Francisco, CA, pp. 323–328, 1996.

[89] Z. Pan and T. Başar, "Backstepping controller design for nonlinear stochastic systems under a risk-sensitive cost criterion," submitted to *SIAM Journal of Control and Optimization*, 1996, also in *Proceedings of the 1997 American Control Conference*, Albuquerque, NM, pp. 1278–1282.

[90] A. A. Pervozvanskii, "Continuous extremum control system in the presence of random noise," *Automation & Remote Control*, vol. 21, pp. 673-677, 1960.

[91] V.M. Popov, *Hyperstability of Automatic Control Systems*, Editura Academiei Republicü Socialiste România, Bucharest, 1966 (in Romanian).

[92] L. Praly and Z. P. Jiang, "Stabilization by output-feedback for systems with ISS inverse dynamics," *Systems & Control Letters*, vol. 21, pp. 19-33, 1993.

[93] L. Praly and Y. Wang, "Stabilization in spite of matched unmodeled dynamics and an equivalent definition of input-to-state stability," *Mathematics of Control, Signals and Systems*, vol. 9, pp. 1-33, 1996.

[94] K. A. Ross, *Elementary Analysis: The Theory of Calculus*, New York: Springer-Verlag, 1980.

[95] T. Runolfsson, "The equivalence between infinite horizon control of stochastic systems with exponential-of-integral performance index and stochastic differential games," *IEEE Transactions on Automatic Control*, vol. 39, pp. 1551-1563, 1994.

[96] R. Sepulchre, M. Janković and P. V. Kokotović, *Constructive Nonlinear Control*, Springer-Verlag, New York, 1997

[97] R. Skelton, T. Iwasaki and K. Grigoriadis, *A Unified Algebraic Approach to Control Design*, London: Taylor and Francis Publishers, 1997.

[98] E. D. Sontag, "A Lyapunov-like characterization of asymptotic controllability," *SIAM Journal of Control and Optimization*, vol. 21, pp. 462-471, 1983.

[99] E. D. Sontag, "A 'universal' construction of Artstein's theorem on nonlinear stabilization," *Systems & Control Letters*, vol. 13, pp. 117-123, 1989.

[100] E. D. Sontag, "Smooth stabilization implies coprime factorization," *IEEE Transactions on Automatic Control*, vol. 34, pp. 435-443, 1989.

[101] E. D. Sontag and Y. Wang, "On characterizations of input-to-state stability property," *Systems & Control Letters*, vol. 24, pp. 351-359, 1995.

[102] P. Soravia, "$\mathcal{H}_\infty$ control of nonlinear systems: differential games and viscosity solutions," *SIAM Journal on Control and Optimization*, vol. 34, pp. 1071-1097, 1996.

[103] J. Sternby, "Extremum control systems: An area for adaptive control?" Preprints of the *Joint American Control Conference*, San Francisco, CA, 1980, WA2-A.

[104] G. Tao and P. V. Kokotović, *Adaptive Control of Systems with Actuator and Sensor Nonlinearities*, New York: John Wiley & Sons, 1996.

[105] A. R. Teel, "A nonlinear small gain theorem for the analysis of control systems with saturation," *IEEE Transactions on Automatic Control*, vol. 41, pp. 1256–1270, 1996.

[106] A. R. Teel and L. Praly, "On output-feedback stabilization for systems with ISS inverse dynamics and uncertainties," *Proceedings of the 32nd IEEE Conference on Decision and Control*, San Antonio, TX, pp. 1942–1947, 1993.

[107] H. S. Tsien, *Engineering Cybernetics*, New York, NY: McGraw-Hill, 1954.

[108] J. Tsinias, "The concept of 'exponential ISS' for stochastic systems and applications to feedback stabilization", preprint, 1997.

[109] A. J. van der Schaft, "On a state space approach to nonlinear $\mathcal{H}_\infty$ control," *Systems & Control Letters*, vol. 16, pp. 1–8, 1991.

[110] A. J. van der Schaft, "$\mathcal{L}_2$-gain analysis of nonlinear systems and nonlinear state feedback $\mathcal{H}_\infty$ control," *IEEE Transactions on Automatic Control*, vol. 37, pp. 770–784, 1992.

[111] A. J. van der Schaft, "Robust stabilization of nonlinear systems via stable kernel representations with $\mathcal{L}_2$-gain bounded uncertainty," *Systems & Control Letters*, vol. 24, pp. 75–81, 1995.

[112] A. J. van der Schaft, *$\mathcal{L}_2$ Gain and Passivity Techniques in Nonlinear Control*, New York: Springer, 1996.

[113] J. L. Willems and J. C. Willems, "Feedback stabilizability for stochastic systems with state and control dependent noise," *Automatica*, vol. 12, pp. 277–283, 1976.

[114] B. Wittenmark and A. Urquhart, "Adaptive extremal control," *Proceedings of the 34th IEEE Conference on Decision and Control*, New Orleans, LA, December 1995, pp. 1639–1644.

[115] W. M. Wonham, "Optimal stationary control of a linear system with state-dependent noise," *SIAM Journal on Control*, vol. 5, pp. 486–500, 1967.

[116] T. Yoshizawa, *Stability Theory by Lyapunov's Second Method*, The Mathematical Society of Japan, 1966.

Index

4th-moment exponential stability, 55

adaptive backstepping, 102, 120, 122
adaptive control, 95
adaptive control for nonlinear systems, 120
adaptive control Lyapunov function (aclf), 97
adaptive LQR, 110
adaptive quadratic tracking, 101, 102, 106, 111
adaptive stochastic backstepping, 121
adaptive tracking, 97
adaptive tracking control Lyapunov function (atclf), 97
asymptotic performance, 129
atclf, 98, 101, 103, 110
averaging, 167, 172

backstepping, 6, 10, 32, 50, 56, 62, 67, 111
Barbalat's lemma, 5
Bolzano-Weierstrass theorem, 135
boundary layer model, 170

center manifold, 148, 179
center manifold theorem, 142, 146
certainty-equivalence, 108, 110
Chebyshev's inequality, 46, 47, 48
Chetaev's theorem, 143
class $\mathcal{K}$, 3
class $\mathcal{K}_\infty$, 3
class $\mathcal{KL}$, 4
clf, 8, 10

control Lyapunov function (see clf)
convergence in probability, 125
convergence in the mean, 125
cost-to-come method, 120
covariance control, 91

differential game, 40, 77, 85, 87, 90
differential geometric control theory, 10
disconnecting adaptation, 157
disturbance attenuation, 21
Doob, 48
Dynkin's formula, 87

extremum control, 172
extremum seeking, 163

forwarding, 10
Frobenius norm, 78

gain assignment, 17
gain margin, 30, 109
global invariant manifolds, 177
globally asymptotically stabilizable in probability, 57
globally asymptotically stable in probability, 46
globally exponentially stable, 4
globally stable in probability, 45
globally uniformly asymptotically stable, 4
globally uniformly stable, 4
gradient, 129

$\mathcal{H}_2$ problem, 85
Hamilton-Jacobi-Bellman equations, 61, 76, 108

191

Hamilton-Jacobi-Isaacs equations, 21, 40, 86
Hammerstein and Wiener models, 172

incremental covariance, 78, 82, 85, 86, 121
infinitesimal generator, 61, 125
input-to-state stability (ISS), 13, 40, 77
input-to-state stabilizability, 40
input-to-state stabilizable, 15, 16, 21, 25
input unmodeled dynamics, 30, 41
invariant manifolds, 146, 177
inverse optimal adaptive tracking, 106, 110
inverse optimal gain assignment, 18, 21, 25, 27
inverse optimal noise-to-state stabilization, 84
inverse optimal stabilization in probability, 58, 61
inverse optimal stochastic gain assignment, 85, 89
inverse optimality, 3
iss-clf, 15, 16, 79
iss-clf-scpj, 26
ISS-control Lyapunov function (see iss-clf)
ISS-Lyapunov function, 14
Itô's differentiation rule, 51, 67, 82, 122
Itô's integral, 59
Itô isometry, 48

LaSalle-Yoshizawa theorem, 4
least excitation, 129, 132
least-squares, 129
Lebesgue integral, 135
Lebesgue measure, 150, 153, 178
Legendre-Fenchel transform, 18, 58, 85, 175
LQG, 90

linearization theorem, 142
Lyapunov theorem, 6

multiplicative noise, 127

noise-to-state stability (NSS), 77, 90
noise-to-state stabilizable, 81
noise-to-state stabilization, 79, 82, 126
nonlinear dynamical systems issues in adaptive systems, 161
nonlinear $\mathcal{H}_\infty$, 13, 17, 40
nss-clf, 80, 81, 84, 89
NSS-control Lyapunov function (see nss-clf)
NSS-Lyapunov function, 79

observer, 66
optimal adaptive control, 120
output feedback, 66, 76, 120
output-feedback systems driven by white noise, 66

partial excitation, 130, 132
persistent excitation, 129, 132
phase margin, 30

quartic Lyapunov functions, 50
quasi-steady state, 170

radially unbounded, 23, 25, 62
risk-sensitive control, 74, 84, 90
robust control Lyapunov functions, 41
robust nonlinear control, 40
reduced model, 170
regressor, 129

sclf, 57, 61
self-optimizing control, 172
singular perturbation, 170
small control property, 9, 10, 16, 26, 57, 61, 81, 84, 98
Sontag's formula, 7, 98, 104

Sontag-type control law, 16, 21, 57, 81, 110
stability, 4, 10
stability margins, 26, 28, 41, 91
stabilization, 6, 8, 10
stable invariant manifolds, 147, 148
stochastic adaptive regulation, 121
stochastic control Lyapunov function (*see* sclf)
stochastic differential games, 85, 90
stochastic ISS, 90
stochastic (linear) adaptive control, 121
stochastic Lyapunov and LaSalle like theorems, 45, 74
strict-feedback systems, 6, 33, 34, 50, 76, 82, 105, 114, 121
strictly passive, 29, 31

terminal penalty, 18
Tikhonov type theorem on the infinite interval, 171
time scales, 166
topologically equivalent, 146
transient performance, 119
tuning function, 100, 104, 122, 131

universal formula, 8
unstable invariant manifold, 148

Viète's formulae, 148

Wiener process, 45, 50, 66, 78, 85, 121
"worst case" disturbance, 20

Young's inequality, 52, 60, 63, 176